CLEANING

For over a decade, the Science for Conservators volumes have been the key basic texts for conservators throughout the world. Scientific concepts are fundamental to the conservation of artefacts of every type, yet many conservators have little or no scientific training. These introductory volumes provide non-scientists with the essential theoretical background to their work.

The Heritage: Care–Preservation–Management programme has been designed to serve the needs of the museum and heritage community worldwide. It publishes books and information services for professional museum and heritage workers, and for all the organizations that service the museum community.

Editor-in-chief: Andrew Wheatcroft

The Development of Costume
Naomi Tarrant

Forward Planning: *A handbook of business, corporate and development planning for museums and galleries*
Edited by Timothy Ambrose and Sue Runyard

The Handbook for Museums
Gary Edson and David Dean

Heritage Gardens: *Care, conservation and management*
Sheena Mackellar Goulty

Heritage and Tourism: *In 'the global village'*
Priscilla Boniface and Peter J. Fowler

The Industrial Heritage: *Managing resources and uses*
Judith Alfrey and Tim Putnam

Managing Quality Cultural Tourism
Priscilla Boniface

Museum Basics
Timothy Ambrose and Crispin Paine

Museum Exhibition: *Theory and practice*
David Dean

Museum, Media, Message
Edited by Eilean Hooper-Greenhill

Museum Security and Protection: *A handbook for cultural heritage institutions*
ICOM and ICMS

Museums 2000: *Politics, people, professionals and profit*
Edited by Patrick J. Boylan

Museums and the Shaping of Knowledge
Eilean Hooper-Greenhill

Museums and their Visitors
Eilean Hooper-Greenhill

Museums without Barriers: *A new deal for disabled people*
Foundation de France and ICOM

The Past in Contemporary Society: *Then/now*
Peter J. Fowler

The Representation of the Past: *Museums and heritage in the post-modern world*
Kevin Walsh

Towards the Museum of the Future: *New European perspectives*
Edited by Roger Miles and Lauro Zavala

Museums: A Place to Work: *Planning museum careers*
Jane R. Glaser and Artemis A. Zenetou

Marketing the Museum
Fiona McLean

Managing Museums and Galleries
Michael A. Fopp

Museum Ethics
Edited by Gary Edson

SCIENCE FOR CONSERVATORS

Volume 2

CLEANING

Conservation Science Teaching Series

MUSEUMS &
GALLERIES
COMMISSION

LONDON AND NEW YORK

ntific Editor
athan Ashley-Smith
per of Conservation
toria & Albert Museum

ies Editor (Books 1–3)
en Wilks

iser
ham Weaver
or Lecturer
artment of Materials
nce
lty of Technology
n University

Authors
Anne Moncrieff
Conservation Officer
Science Museum

Graham Weaver
Senior Lecturer
Department of Materials Science
Faculty of Technology
Open University

Advisers
Jim Black
Summer Schools
Institute of Archaeology
University College London

Suzanne Keene
Head of Collections Services Group
Science Museum

Jane McAusland
Private Conservator

Anna Plowden
Private Conservator

---

published by the Crafts Council 1983
nd impression 1984

ished by The Conservation Unit of the
eums & Galleries Commission in 1987

hardback and paperback edition published in 1992
outledge
k Square, Milton Park, Abingdon, Oxon, OX14 4RN

ltaneously published in the USA and Canada
outledge
hird Avenue, New York, NY 10017

edge is an imprint of the Taylor & Francis Group, an informa business

rations by Berry/Fallon Design
ned by Robert Updegraff and Gillian Crossley-Holland

Library Cataloguing in Publication Data
alogue record for this book is available
the British Library

y of Congress Cataloguing in Publication Data
alogue record for this book is available
the Library of Congress

0–415–07165–8

Contents

eface to the 1992 edition

: science of conserving artworks and other items of cultural significance has lergone considerable change since 1982 when this series was instigated, stly involving the development or application of new materials or nniques. Their understanding by conservators, restorers and students tinues, nonetheless, to depend on familiarity with the underlying scientific nciples which do not change and which are clearly explained in these books.

n response to continued international demand for this series, The nservation Unit is pleased to be associated with Routledge in presenting se new editions as part of The Heritage: Care–Preservation–Management gramme. The volumes are now enhanced by lists of recommended reading ch will lead the reader to further helpful texts, developing scientific ideas conservation setting and bringing their application up to date.

Introduction

Alice thought she had never seen such a curious croquet ground in all her life; it was all ridges and furrows; the balls were live hedgehogs, the mallets live flamingoes, and the soldiers had to double themselves up and to stand upon their hands and feet, to make the arches.

The chief difficulty Alice found at first was in managing her flamingo: she succeeded in getting its body tucked away, comfortably enough, under her arm, with its legs hanging down, but generally, just as she had got its neck nicely straightened out, and was going to give the hedgehog a blow with its head, it *would* twist itself round and look up in her face, with such a puzzled expression that she could not help bursting out laughing: and when she had got its head down, and was going to begin again, it was very provoking to find that the hedgehog had unrolled itself, and was in the act of crawling away: besides all this, there was generally a ridge or a furrow in the way wherever she wanted to send the hedgehog to, and, as the doubled-up soldiers were always getting up and walking off to other parts of the ground, Alice soon came to the conclusion that it was a very difficult game indeed.

Alice's Adventures in Wonderland
Lewis Carroll, 1865

Cleaning treatments used in conservation are, of course, no game but many of the problems Alice has to cope with are analogous to the conservator's. Essentially, all cleaning might be described as hitting away (using many different methods and on widely differing scales) unwanted material from an object that is often flawed and fragile. Although the reasons for cleaning can vary, once the decision to clean has been taken, the conservator's problems are fundamentally ones of *control*. It is here that this book (and the others in the series) will help you. By learning the scientific principles governing the cleaning methods you use, it is possible to systematise and understand more fully what controls their effectiveness and applicability.

As with all the books in the series, this, the second of six, has been designed for conservators working on all kinds of material, and there is much scientific inter-relation between many of the methods used. Do remember, however, that this book is not intended as a conservation handbook or manual and will not help you to solve all the ethical questions of whether or not to clean. What it can do, however, is to inform you of the possible risks and the extent of damage that might be caused through using one method rather than another.

The Crafts Council and the team who have worked on these books hope that their publication will provide you with a useful scientific base which will enable you to go on to read with some confidence the specialist papers and books on conservation which are already available. It is hoped, too, that the series as a whole will form a useful text for conservation courses.

Using This Book

As the second book in the series, it is assumed from Chapter 1 onwards that you will have already read and become familiar with the contents of Book 1, *An Introduction to Materials*. This means that you will now think of science as a systematic and structured discipline and will have an understanding of basic chemistry. It is assumed, therefore, that you will be able to read chemical names, structural formulae and chemical equations; that you are familiar with atomic structure and chemical bonding, and hence how chemical structure relates to the physical and chemical properties displayed by some common materials used or worked on by conservators.

Book 2 covers much of the science involved in a wide variety of cleaning procedures, both mechanical and chemical. It explains what holds dirt in place and discusses, amongst other things, the science underlying the properties of liquids and the use of solvents. It includes an explanation of pH and introduces the scientific concepts of energy and colour.

Because these books have been specially prepared for practising conservators, some readers may find that the order in which the science is taught varies quite considerably from other more standard textbooks. It is especially important, therefore, that you read this book in its proper sequence. Remember that Books 1 and 2 will not form a complete scientific syllabus in themselves; you will need to go on to read Book 3 (where you will learn far more about polymers) and the subsequent books before a comprehensive syllabus is built up.

Do not try to read this book too quickly, often the science contains ideas which take time to absorb properly. Ask someone with a scientific training to help you if any part still seems obscure after several readings. Some simple demonstrations have been included to illustrate or clarify certain points in the text. New scientific words and terms are printed in bold type and are repeated in the outer margins for easy reference. Cross-references are also made at certain points in the text where it may be useful for you to refer back to a previous passage (or to Book 1). A full index is included at the end of the book.

Acknowledgements
This series is being prepared by a team of conservation scientists, conservators and science teachers. The Crafts Council is deeply grateful to the conservators and in particular the conservation scientist, who, as the joint author of this book has given an enormous amount of her own time to its preparation over the last three years. The Council also wishes to acknowledge the generosity of the institutions and private workshops (in particular the Victoria & Albert Museum and the Open University) who have lent their support through allowing their staff to work with us. The contributions made to such a complex and difficult educational task have been necessarily varied, but each has been of great value and importance. The Council is especially indebted to Jonathan Ashley-Smith for his contribution as scientific editor.

October 1982

1

The nature of dirt

A What dirt is and why you clean
B What holds dirt in place?
C Chemically bonded dirt
D Cleaning techniques

The nature of dirt

This chapter deals with some of the general features of dirt and points to some of the factors that need to be considered when carrying out cleaning in conservation work. Physical and chemical reasons for the adherence of dirt to objects are discussed and the basic principles behind cleaning techniques are outlined.

A What dirt is and why you clean

Dirt can be defined as material which is in the wrong place, rather as a weed is thought of as a plant growing in the wrong place. The evidence of blood on textiles or the remains of food in a vessel may be considered "dirt" on modern objects but may have to be carefully preserved on ancient artefacts. As a conservator, you often need to be able to remove material which is in the wrong place (for a variety of reasons) without removing material which is in the right place. This process is frequently complicated by the fact that the substance of the object may be very similar to the dirt.

A major objective of all conservation treatment is to increase the chemical stability of the object being treated. **Cleaning** often forms an important part of the stabilizing process. This is because dirt on an object can be a potent source of deterioration (as, for example, when chloride salts set up corrosion reactions on bronze, or moulds grow on organic materials like paper or textiles). At other times, cleaning may be a necessary preliminary to a further treatment, as when preparing a surface before coating or joining. In many instances cleaning requires delicate judgement and experience on

cleaning

the part of the conservator, in deciding what the final appearance of the object being cleaned should be and how much "dirt" should be kept (patina, historical evidence, etc). It isn't easy to choose a cleaning method when you want to remove only part of the dirt.

It is helpful to begin by classifying dirt into two categories:

Foreign matter which is not part of the original object.
Examples: soot, grease, stains, adhesives and fillings from old repairs.

Products of alteration of the original material of the object.
Examples: metal corrosion products, yellowed varnish, decayed timber or stone.

foreign matter

product of alteration

The scientific basis of this classification of dirt is to be seen in the distinction between physical mixtures and chemical compounds drawn in Book I (Chapter 2). Dirt which is **foreign matter** was not originally present in the substance of the object but has later become *mixed* with it. In contrast, dirt which is a **product of alteration** has formed through a *chemical combination* of the original material with chemicals from the environment such as gases in the air or salts in solution from soil or the sea.

dust

"Foreign matter", however, can give rise to "a product of alteration" if a chemical reaction occurs between the dirt and the object. Nevertheless, even if nothing so harmful as this does occur, it is unwise to assume that loose foreign matter (**dust**) is innocuous or that it can be safely left as a thick layer on objects in store. Dust is commonly an amazing mixture of fragments of human skin, textile fibres, carbon particles (soot), and grease from unburned hydrocarbon fuels, from cooking and from the skin of people and animals. There are often many salts in dust, for example, sodium chloride (carried in from sea spray or on skin fragments), and sharp gritty silica crystals are often present. In this chemical mixture are the spores of countless moulds and fungi and micro-organisms which live on the organic material in the dust. These organisms are equally likely to attack *objects* made of organic material. Much of this dirt is *hygroscopic* (water-attracting) and this tendency can encourage the growth of moulds and increase the corrosiveness of salts. So even dust is damaging, although perhaps only slowly.

If the dirt is a product of the alteration of part of an object you can see immediately that by removing it you are *actually taking away some of the object itself*. Thus the tarnish on silver is black silver sulphide formed by reactions of the silver in the object with hydrogen sulphide and moisture from the air. When you clean away the tarnish you not only remove the sulphur atoms, which are foreign matter, but also some silver atoms originally positioned by the silversmith. Hence, "clean" though it may be, the object is less than it was and the surface is not the original one. Perhaps the tarnish should have been left on? Perhaps it would be better to reverse the chemical reaction to recover the silver metal? Science cannot answer the ethical question whether the tarnish ought to be removed, but it may offer methods which avoid removal *and* get a clean surface. It gives you options beyond "take it or leave it".

To remove dirt which is a product of deterioration *necessarily* involves taking away some of the original artefact. In principle, removing dirt composed of foreign matter does not imply this. In practice, however, it is rarely possible to separate the dirt from the object without taking some of the object too. The difficulties derive from the natural or corrosion-induced porosity of the surface and the extreme fineness of the dirt as it starts to coat the object. Soot particles in smoke may be as small as one micron (μ) in diameter ($\frac{1}{1000}^{th}$ of a millimetre) and will penetrate the finest crevices of a surface. In solution the foreign matter can reach an atomic scale of fineness, and it will be deposited when a solvent being used for cleaning evaporates. To remove *only* the dirt when it is so intimately mixed with the object is not easy and becomes more and more difficult the cleaner you want the final product to be. The penetration of crevices in a fragile porous surface during the cleaning process can easily cause the thin walls between the pores to break down. Some degree of damage to the object itself is likely, as illustrated in Figure 1.1.

Figure 1.1 *Different levels of cleanness showing how damage can occur through over-cleaning.*

There are many questions that need to be considered when undertaking or deciding on a cleaning treatment. Some have already been raised in describing the nature of dirt itself but there are many others which relate to questions of conservation ethics and historical study. Science cannot take from you the responsibility of making these judgements but it can explain the actions and consequences of a wide variety of cleaning treatments which will help to *inform* your decisions. Faced with a dirty object you need to ask (and answer) the following questions:

Why clean?
Is it dirt? Should some or all of it remain?
Is the dirt doing damage?

Can the object tolerate being cleaned?
What are the physical and chemical properties of (a) the object; (b) the dirt?
What will affect the dirt without affecting the object?

What will be the effect of cleaning?
What will be the appearance of the object after cleaning?
Will the stability of the object be affected?
How often will the object need cleaning in future?

How can you clean the object?
Is there a suitable treatment?
How does the treatment work?
Is the treatment safe both for you and the object?
When do you stop?

B What holds dirt in place?

Dust may adhere only lightly to a surface or may become mechanically trapped in the interstices of a porous solid (as shown in Figure 1.1) or, even more obviously, may get trapped in fibrous material such as textiles. The dirt may simply be entangled in fibres, or particles may be locked into crevices like pieces of a jigsaw fitting together. Dust may also be held on a surface by **electrostatic attraction**: the generation of *static electricity* by friction was described in Book I (Chapter 4). When two materials are rubbed together the friction forces at the point of contact can knock loosely held electrons off one surface onto another. The surface which gains electrons is said to be **negatively charged** and the one that has lost them becomes **positively charged**. Which surface becomes positively or negatively charged depends upon what is rubbed with what, but both types of charged surface act equally as dust collectors. When you polish a surface, mobile charged particles (electrons) are rubbed off the polishing cloth and deposited onto the surface being polished.

electrostatic attraction

negative and positive charges

If you are going to deal with the problem it is useful to know:

- How the dust is attracted
- Why the electric charge stays put, and
- What can be done to disperse or avoid a build-up of static charge.

Charged surfaces attract dust because the dust particles themselves contain electrons which can be displaced. Figure 1.2 illustrates how this happens.

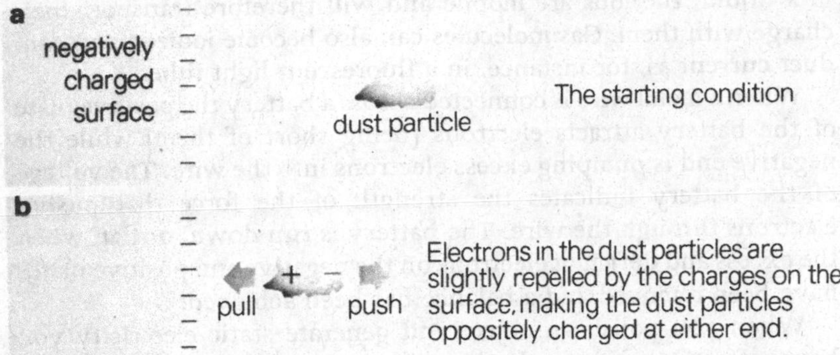

Figure 1.2

The attraction between the surface and the + end of the dust particle is stronger than the repulsion between the − end and the surface because the + end is nearer.

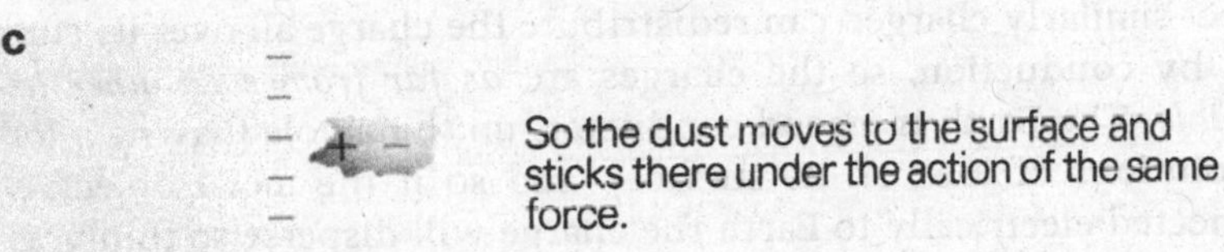

c
So the dust moves to the surface and sticks there under the action of the same force.

Figure 1.2

Once an electrically charged surface has been produced there is no way of preventing the dust moving on to it. Attempts to remove the dust by wiping will not work because *more* friction is created in the process and reinforces the static charge. The dust will jump straight back from the duster to the charged surface. The answer is to get rid of the charge, or if possible not to generate it at all. To understand these manoeuvres you need to know more about the contrasting electrical properties of materials; the consequences of their being either **conductors** or **insulators**.

conductors and insulators

It is a familiar fact that metals conduct electricity while most other materials do not. This observation is one of the pieces of evidence upon which the models of atomic bonding are based. In metals some electrons are shared among all atoms and are readily able to move throughout the solid (see Book I, Chapter 4).

The passage of electrons constitutes an electric current. Therefore, solid material in which the electrons are tied up to one ion or molecule (through ionic or covalent bonding) does not allow this easy transfer of electrons. As there is no movement of charge, no current is conducted. Such materials are termed *insulators*. Remember that electrons are not the only agents by which an electric charge can move around. Ions can also carry charge if they are made free. As you know, ions congregate into crystalline solids with + and − ions alternating in the structure. If such a solid is melted or dissolved in a liquid, the ions are mobile and will therefore transport their charge with them. Gas molecules can also become ionized and conduct current as, for instance, in a fluorescent light tube.

When a metal wire is connected across a battery the positive plate of the battery attracts electrons (being short of them) while the negative end is pumping excess electrons into the wire. The voltage of the battery indicates the strength of the force that pushes electrons through the wire. The battery is run down, or flat, when the excess and deficit of electrons on the negative and positive plates have been removed and a balance has been achieved.

When you polish an object and generate static electricity you have effectively charged up a battery. The object forms the equivalent of a negative plate (with excess electrons) and the polishing cloth is the positive plate. What happens after you take the cloth away depends on what electrical connections are made with the charged object.

Both insulators and conductors can become charged. If an insulator (a Perspex display case, for example) has been charged by polishing the charge will stay exactly where it is. However, a metal object similarly charged can redistribute the charge all over its surface by conduction, so the charges are *as far from each other as possible*. The Earth is a good conductor, on the whole (having a lot of solutions of ions on its surface), and so if the metal object is connected electrically to Earth the charge will disperse so thinly as to be unnoticeable and the object is described as having been

earthed charges

earthed. Where the metal is *insulated* from Earth the charge remains on the metal.

Though static charge on metal is easily diffused by conduction, that on insulators is not. The difficulty comes from the inability of the static charge to move across the surface. This means that *every* point of the object needs to be earthed, not just one. So, to be effective, a conductor has to cover the entire surface. Obviously, in the case of a Perspex display case, placing a sheet of metal over its entire surface is not on, so another means of dispersing the static charge must be found.

You are probably aware that static is less of a problem in humid conditions. This is because a layer of water (if only one molecule thick) on a surface connecting it to the earth is sufficient to conduct away the static charge. Although *pure* water is not good electrical conductor, a thin layer of water exposed to the air on one side and touching an object on the other will not be pure, and will contain enough ions to conduct away the offending charge.

C Chemically bonded dirt

The mechanisms of bonding between atoms to form molecules were dealt with in detail in Book I (Chapter 4). Three main types were identified, characterised by electrons being exchanged between atoms (ionic), shared between two atoms (covalent), or shared among many atoms (metallic). Collectively these are known as **primary bonds** and all produce strong forces holding atoms together to form molecules or crystals. Chemical reactions involve changes in the patterns of primary bonds between atoms. However, there are forces which hold *molecules* together. These forces are called **secondary bonds**, and as the term implies, they represent much weaker forces than primary bonds. It is these bonds which explain how materials can stick together without reacting and in particular how dirt sticks to objects. You will remember that the demarcation between ionic bonds and covalent bonds is not sharp. In particular, atoms such as oxygen which receive electrons in making ionic bonds will also make covalent bonds in which they get the greater share of the electrons. The negative charge of the electrons is displaced relative to the positive charge on the nucleus. The distribution of charges in the molecule is not even. Most kinds of secondary bonds have their origin in uneven charge distribution.

primary bonds

secondary bonds

1 Dipole forces

A covalently bonded molecule containing oxygen will have its electrons distributed so that the oxygen atom is slightly negatively charged. Since the whole molecule is electrically neutral (zero charge), some other part must be slightly positive (Figure 1.3).

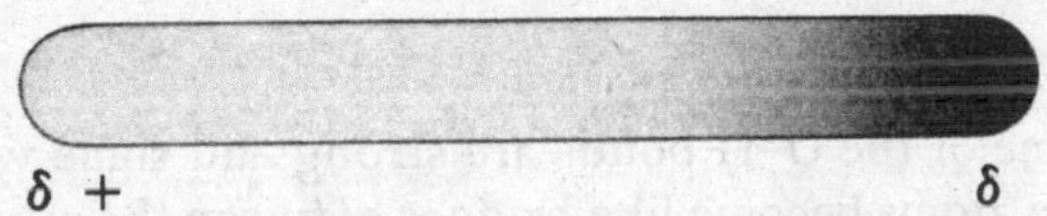

Figure 1.3 *The positive and negative charge distribution on an electric dipole.*

You will see that the imbalance of charge is shown in this diagram as $\delta+$ and $\delta-$. The Greek letter delta, δ, is conventionally used to denote "a little bit", in this case rather less than a whole electron's-worth of charge. Molecules charged like this are known as electric **dipoles** (*di* = two, *poles* = ends). A substance is called *polar* if its molecules are dipoles and *non-polar* if they are not. Dipolar molecules behave rather like bar magnets, unlike ends sticking together in patterned arrays, as shown in Figure 1.4.

dipoles

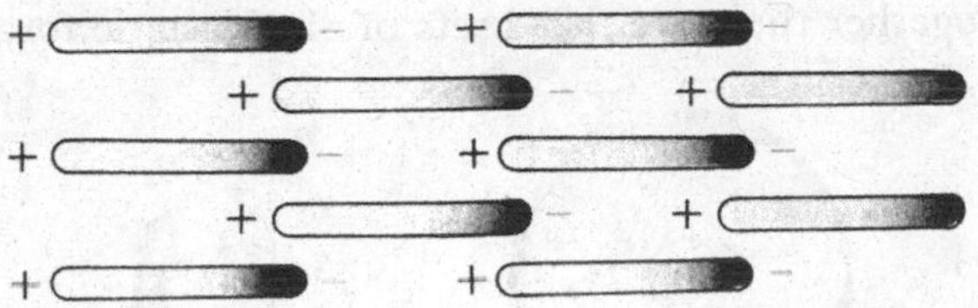

Figure 1.4 *An array of dipolar molecules illustrating schematically the distributions of positive and negative charge.*

This arrangement is similar to those of ionic crystals (see Book I, Chapter 4) where the charge involved has the strength of at least one whole electron and the shapes of the ions allow tight packing. The electrostatic forces in ionic crystals can thus be very strong. Dipolar molecules, on the other hand, with a smaller charge which is often on an irregular shape, can interact only weakly. This weak attraction is present in the solid, liquid and gaseous states and is one of the causes of the noticeable differences between **polar and non-polar materials.**

polar and non-polar materials

2 Hydrogen bonds

A stronger type of dipole bond occurs when the atom from which oxygen is drawing electrons is a hydrogen atom. Water is the classic case. The hydrogen atoms are slightly positive and are therefore attracted by the negative charge on the oxygen atoms in neighbouring molecules. The molecules tend to form a regular array (Figure 1.5)

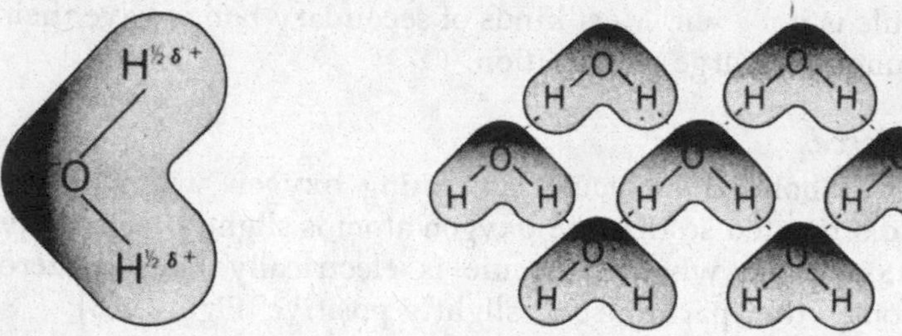

Figure 1.5

in which some of the O–H bonds are strong and some weaker, and the hydrogen atoms become like bridges between the oxygen atoms. When hydrogen atoms make bridges in this way it is called **hydrogen bonding.** Hydrogen bonding is common (because water is common) and accounts for a host of phenomena of importance in conservation, which will be dealt with in this book and in Book 3. Thus, for example, cellulose fibres in cotton and linen textiles and in wood are very strong. This is because there are hydrogen bonds between the many –OH groups on the cellulose molecules. (See Book I, Chapter 4 for the structure of cellulose.)

hydrogen bonding

3 Van der Waals bonds

The final, and weakest, type of secondary bond to be discussed here was first explained by the scientist Van der Waals. When two atoms are close together there are three sets of electrostatic forces acting:

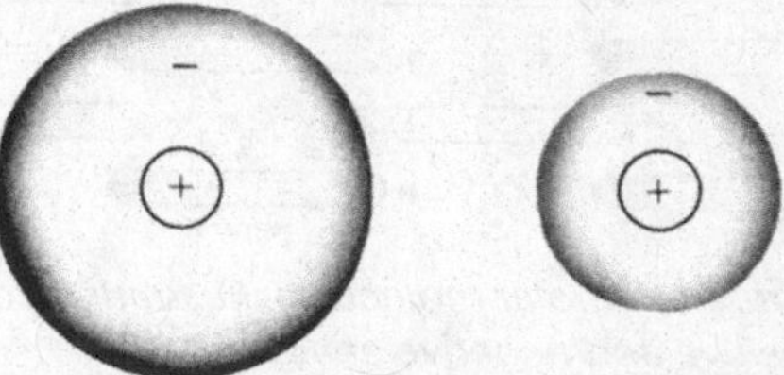

Figure 1.6 *There are attractive forces between the positive and negative areas of these two atoms.*

- The positive nuclei repel one another.
- The two clouds of electrons repel one another.
- Each nucleus *attracts* the electrons of the other atom.

Figure 1.2 showed how an imbalance of charge was induced in a particle of dust, causing it to be attracted to a charged surface. The forces acting on the electron cloud of one atom tend to distort it and so induce an imbalance of charge (dipole) in the atom, if only for a short time. Such dipoles are constantly appearing and disappearing but there is an overall attractive force between the induced dipoles. This means that there is a *weak attractive force acting between all atoms and molecules at all times*. The emphasised words are really all you need to know about **Van der Waals forces.** It is important that these forces act between *all* atoms because this means that molecules will stick together, even when there are no attractive forces due to permanent charge separation in the molecules themselves. For example, paraffin wax is held together by these forces and many things will in turn stick to wax. As wax will stick on the surface of objects, dust sticks in the wax producing typical "foreign matter" dirt. Chapter 3 shows just how solvents work in terms of these forces. The slippery nature of most waxes shows that their molecules are not held together strongly, in contrast with crystalline solids like sugar (hydrogen bonded) and salt (ionic bonds) where the electrostatic forces are much stronger.

Van der Waals forces

Generally speaking, dirt is held to the surface of objects by one or other of these secondary bonds. The problems of cleaning can be usefully looked at as comparisons of the strength of the **adhesion** of the dirt to the object and the **cohesion** of the molecules of the object to one another. Obviously dirt weakly stuck to a strong object is going to be easy to clean off without damaging the object (unless the object is porous and dirt is trapped in pores and crevices), while dirt firmly stuck to a weak object presents difficult problems. In other words, as you will know from practical experience, when cleaning an object you have to consider not only the dirt you want to remove but also the material you want to take it from. Where the bondings are similar the task is awkward, and where the forces of adhesion are stronger than those of cohesion the problem is difficult.

adhesion and cohesion

D Cleaning techniques

It is because there is so much variety in the combinations of adhesion and cohesion of dirt, objects and cleaning solutions that it is necessary for you to use widely varied techniques. A method has to be chosen so that you can attack the dirt without attacking the object. Sometimes mechanical methods can be used, mechanical force being used to break the bond between the dirt and the object. In a shipyard, for example, a pneumatic hammer might be used to remove rust. This quite obviously constitutes a brutal attack (though in this case, justified by the scale of the job). However, all the other, far more gentle approaches to breaking the bonds between dirt and object can also be looked upon as some kind of hammering, and doing so brings out some common features of cleaning techniques and especially highlights the inevitability of damage during cleaning. A dirty object is one on to which extra material has moved; cleaning is the business of *re*-moving that extra material. One class of cleaning techniques, those called **mechanical**, have in common the method of getting the dirt to move. All involve arranging a collision between the dirt and some material object. As an analogy think of a sculptor who uses a variety of tools to remove the superfluous stone (dirt) to reveal the desired form (object). Points and chisels of increasing fineness are made to collide with the surface and it is hoped that, with each blow, nothing more than the unwanted material will be struck off. When cleaning an object consider the following questions. Can you arrange only to hit the dirt? Answer – no. Will hitting the object damage it? Answer – yes. Will the dirt part from the object cleanly? Answer – maybe. Where will the dirt go to? Answer – back on to the object unless you collect it.

mechanical cleaning techniques

Chemical cleaning is not exempt from the hammering analogy. In Book I you learnt how chemical bonds are broken by impacts from adjacent molecules. At ordinary temperatures the velocities of molecules are of the order of several thousands of kilometres per hour. Any solvent or reactant you put on an object and its dirt causes chemical collisions with both, and although the dirt may come off you can guarantee that some traces of the chemical you have used will have worked their way into the object. Again, minute damage – but it is there. The useful thing about using chemical collisions is that they can be chosen to be highly selective; you can arrange to do far greater damage to the dirt than to the object, provided you know enough chemistry to make a wise choise of cleaning substance.

It should be obvious that gentleness does not so much prevent damage as render it invisible. Damage, to some degree, occurs *whatever* method is used.

2

Mechanical cleaning

A Review of mechanical cleaning techniques

- *A1 Dusting and polishing*
- *A2 Picking methods for removing solid layers*
- *A3 Abrasive methods*
- *A4 Vibrations*

B Energy: familiar word – new concept

Mechanical cleaning

In the last chapter we looked at the nature of dirt and the ways it can adhere to an object. This chapter concentrates on the science affecting the various forms of mechanical cleaning which conservators use. The concept of energy as it relates to mechanical cleaning processes is introduced.

A Review of mechanical cleaning techniques

All the cleaning techniques termed "mechanical" involve arranging a collision between the dirt and some material object. The purpose of the collision is to produce a force which will (a) break the contact between dirt and object and (b) move the dirt away from the object. You have to try and ensure that the break occurs at the interface between object and dirt and that the method you use to produce this break actually hits the dirt and not the object. Both these problems are essentially ones of *control*.

Mechanical cleaning, merely breaking the adhesion of the dirt and moving it away, contrasts with chemical cleaning, accomplished either by dissolving the dirt (see Chapters 4 and 5) or by causing something to react with it (Chapter 7). Mechanical cleaning methods have certain advantages for the object since nothing is added during them that might cause further deterioration, such as solvents carrying the dirt further into porous materials, or water causing swelling in hygroscopic (water-attracting) materials or starting corrosion in metals. Similarly, for the conservator there is no involvement with toxic chemicals. You must note, however, that virtually all mechanical cleaning processes require protection for the eyes, skin and lungs of the operator. You should take care to consult safety manuals and the instructions for the use of equipment before undertaking work of this kind.

safety

Questions that must be answered when choosing a mechanical cleaning method are:

- How firm is the adhesion between dirt and object?
- Is the dirt brittle (easily fractured) or tough (resistant to fracture)?
- What are the mechanical properties of the object?

So complex are the inter-relations of these factors, even within a single object, that specific guidance is difficult and would be out of place in a book like this. What can be done is to show the influence of general scientific factors upon each treatment as it is reviewed. Another scientific idea needs to be introduced. This is the concept of **energy** which is sometimes defined as the ability to do work; the "work" of interest here being mechanical cleaning.

energy

A1 Dusting and polishing

Dust is dirt which has not *co*hered to itself nor *ad*hered very strongly to the object. It can therefore easily be moved with a cloth or a brush or a feather duster. Delicate and gentle the process may be, but you cannot ensure that only the dust is moved. Threads or feathers pulling at flakes on the surface of an object can move these too and you may have seen fragments of cloth or feathers caught in cracks in objects. Those fragments of torn off duster represent a gamble being taken – that the feathers will break before the object does. Also, once you have moved the dust, where does it go to?

A vacuum cleaner is another kind of duster. At least it has the advantage that most of the dust shifted off the objects is contained in the bag. Vacuum cleaners move dust by causing a rush of air through and past the object being cleaned. A stream of air molecules strikes the dust and makes it move – together perhaps with loose parts of the object. One snag is that the amount of suction provided by a machine is largely pre-determined by the manufacturer. You can exert some control on how much air goes through or past the object by using different attachments or by holding the nozzle closer to or more distant from the object. The wind speed can also be controlled by the size of the nozzle, narrow ones producing a higher speed flow.

When dirt has been put in motion, damage may be caused for other reasons. If the dust is harder than the material of the object itself, it may, once moving, scratch or cut into the object's surface. Grit causes damage when it is moving. This is why the sharp sandy dust from people's shoes must be removed from carpets. Even with a drugget or other protective material on top, the grinding action of walking can cause damage which can be measured by suitable test equipment. The cleaning process also does damage, but to a lesser extent.

Polishing is another common treatment for certain objects. Although sparkling glass and glossy furniture may be pleasing to look at, polishing can cause harm to the object. There are three possible reasons for this:

1 Abrasion

The first problem with polishing is **abrasion** (scratching). The cause is the same as it is when dusting – hard particles on the move will cut into soft surfaces. Some polishes (for example those for glass and metal) deliberately contain abrasive particles, the intention being to cut away thin films of well-bonded dirt or tarnish. Inevitably such a cleaning compound will produce microscopic scratches on the surface of the object too. If such compounds are used with care and knowledge, a minimal level of abrasion may be considered acceptable in some cases. Much worse, however, is damage caused by foreign particles of larger size which may find their way on to your polishing cloth. There is only one way to avoid this: scrupulous cleanliness of the object, the cloth and the polish.

abrasion

2 Static electricity

Friction as a cause of static electricity was described in the previous chapter (Section B). Even the friction of dusting *may* be enough to make a surface attract dust again but polishing is *always* sufficient. An electrically charged surface will attract dust, which cannot be wiped away. First, the additional friction reinforces the **static charge**; second, the dust will jump straight back from your duster to the charged object. The cure must be sought in getting rid of the charge or never generating it in the first place. There are several ways of doing this:

static charge

- **Conducting the charge away directly.** As you will remember, this is difficult to do on an insulator since every point on the surface needs to be connected to the earth and this is not possible, in practice, with wood, glass or plastic.
- **Conducting the charge away indirectly** by providing charged ions over the whole surface. Water molecules will do this if the humidity of the room is raised, dispersing the charge away from the surface. The ions left behind by antistatic polishes may also help by attracting water molecules to the surface of the object. The charge on an object can be neutralised or conducted away by charged ions in the air. These can be generated in several ways. Ultra-violet light will ionize the air and the ions will be attracted to a surface with the opposite charge, but of course even small quantities of ultra-violet light are not wanted where organic

materials are on display, because of the deterioration it causes. Commercially available electrostatic "guns" produce charged ions and are sometimes effective at neutralising the static build-up.

- **Never generating the charge in the first place.** This ideal can be partially achieved by using a vacuum cleaner instead of a duster (the air friction is less than cloth friction although some static may still be created) or by using a polish which reduces the friction of the dusting. All polishes, however, have the disadvantages described below.

3 Wax polishes

The many proprietary waxes and polishes for furniture, glass, etc. all give their effect of glossiness by leaving a very thin layer of a greasy substance on the surface (however subtle a "blend of silicones and natural products" the container proclaims). A dull appearance is caused by surface irregularities too fine for the naked eye to see, which scatter the light in all directions so that no clear reflected image is seen. Polishes fill in the minute dents and scratches to produce the desired smooth finish, which reflects the light uniformly. As well as looking good, a waxed object may be protected somewhat from moisture, which could be beneficial. However, waxing has the disadvantage of making surfaces sticky and attractive to dust. Moreover, unless the polish is specially formulated for anti-static properties, the surface will be electrically insulating (and hence more prone to hold charges generated by friction) and thus attract and hold even more dust. Lanolin impregnation of textiles or leather tends to produce the same qualities and consequences. You may eventually need to use a solvent to remove dirt stuck on by the grease from polishes, so it is worth thinking carefully about the pros and cons of any treatment of this kind which may encourage more dirt to stick.

A2 Picking methods for removing solid layers

This form of mechanical cleaning exploits the difference in physical properties between the dirt, whether corrosion or deposited layer, and the object. It is most often used to break up a hard, brittle crust piece by piece, and remove it from the surface of the object. Splitting off corrosion layers with scalpels, needles and dental picks and

pinging and picking

pinging or **picking** off small flakes is a method often used on porous corrosion layers. On dense compacted layers the point of the tool should be held vertical and close to the edge of the corrosion layer so that only a small piece is broken off at a time. Trying to break off too much at once may break the object as well.

Figure 2.1 shows what you are trying to achieve. First the blade or needle has to fracture the layer through the thickness of the corrosion, then the plane of fracture needs to turn suddenly and proceed along the interface between "dirt" and object.

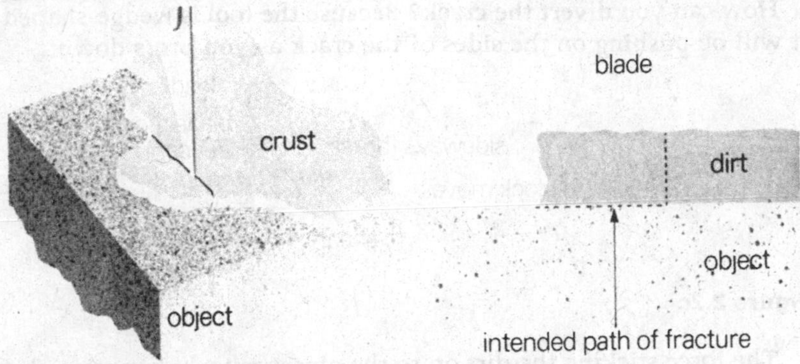

Figure 2.1 *The desired procedure for removing hard brittle crust from an object a small piece at a time.*

Consider first the action of making a crack through the thickness of the "dirt" layer and suppose the tool used is hard enough to make an impression in the surface of the layer. Figure 2.2 (a) shows this condition.

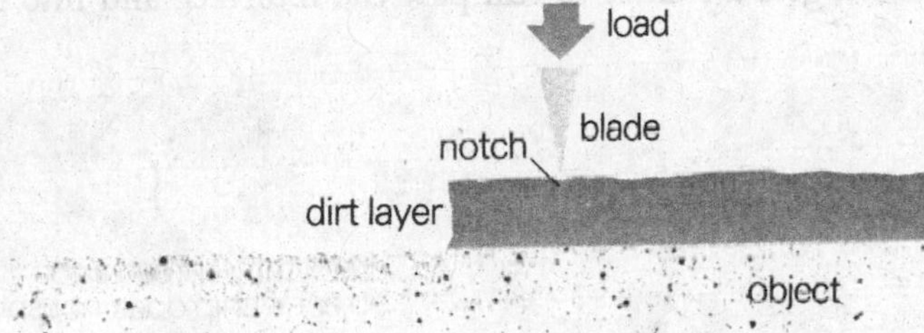

Figure 2.2a

The question now is what will happen at the root of the crack as the wedge-shaped tool tries to open it further. The answer depends on a mechanical property of the material known as its **fracture toughness**. If a layer is brittle, not tough, the crack will run on ahead of the blade down to the interface with the object. If the material is too tough or the layer too thin, the knife blade will reach the object before the crack runs on, causing damage to the object surface. This is why the pinging method only works with brittle dirt of reasonable thickness – you cannot "ping" the silver sulphide tarnish away from silver, for instance, because although the layer is brittle, it is too thin to allow the crack to run ahead of the blade. What should happen is illustrated in Figure 2.2 (b) below:

fracture toughness

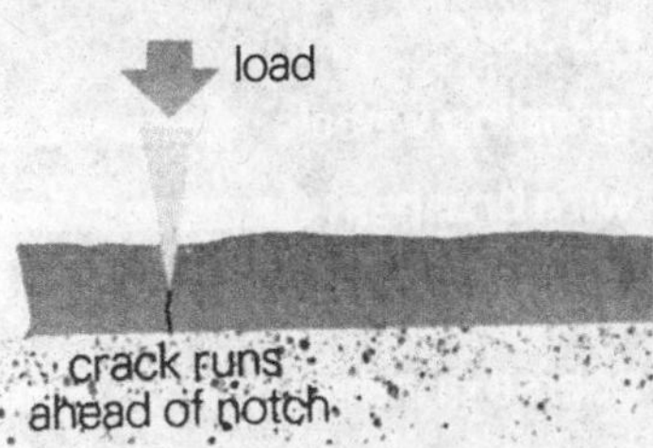

Figure 2.2b

How can you divert the crack? Because the tool is wedge-shaped it will be pushing on the sides of the crack as you press down:

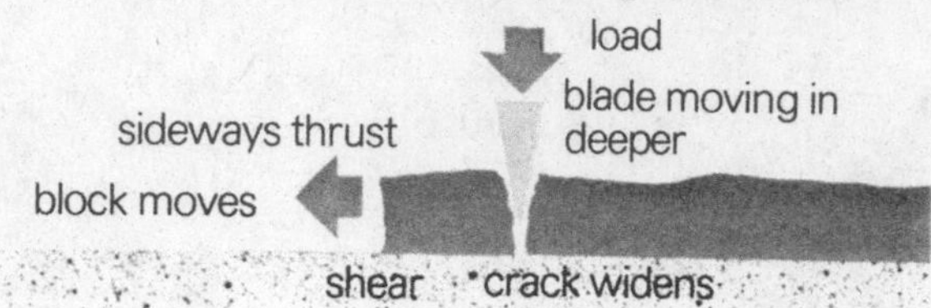

Figure 2.2c

The force sticking the dirt on to the object must be overcome by the sideways thrust from the wedge. Clearly the total force needed to push the chip of dirt sideways (to **shear** it) depends on the strength of the bonding *and* on the area of the chip. This is why you must always go for a *small* chip, where the total force needed will be small.

shear

Taking small chips ensures that dirt shears off *before* the first crack gets a chance to run past the interface and into the object:

Figure 2.2d

Scientific understanding of mechanical processes such as the one just described depends on careful distinction between several mechanical qualities – toughness contrasted with brittleness; strength versus weakness; hardness compared with softness. These matters will be taken up in some detail in Book 3.

A3 Abrasive methods

Abrasion consists of *cutting away* a deposit as opposed to the previous method (snapping it off). Very small fragments of material are cut away and the attack is mounted by many cutting edges simultaneously. The tools come in several guises and range from big and rough to little and gentle:

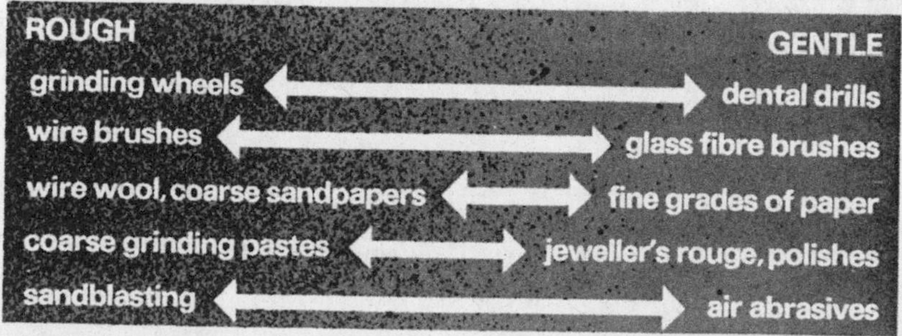

The cutting edges, as you see, are frequently hard, sharp, gritty particles which may be fixed to a rotating disc, glued to paper or cloth, used as a slurry or carried in a blast of air. On impact with dirt or object, indiscriminately, the sharp edges hack away a tiny fragment of the target. The method involves problems of control – it is very important to decide where to stop, especially when the surface of the object is easily abraded.

There is a critical difference between scratching and polishing in the mind of the conservator, the curator and indeed any other observer, but there is no scientific distinction between them. In the end the difference comes down to the size of the scratches – on a polished object they are too fine to be seen. Working with abrasive methods thus requires us to consider the factors which decide whether the object is in danger of being "scratched" in the final stages of cleaning. The relative hardness of the abrasive grit and the object is one factor. Hardness is compared by determining what will scratch what; if the grit is less hard than the object then it cannot scratch it. But the object may not be uniformly hard and the abrasive may be variable, too. Also, it is the hardness of the *crust* that is important until you get through it and contact the surface of the object. If the crust is softer than the object, there is no danger, but corrosion layers and deposits can be very hard. Consequently abrasive grits are chosen for their hardness; examples are sand, carborundum and aluminium oxide. Another factor is the size of the particles of abrasive. Often the same abrasive can be obtained in different sizes of particles, the coarser for preliminary cutting work, the finer for finishing and polishing. Heat results from friction between surfaces during abrasive processes. Water, solvents and oils are sometimes used to remove this heat and to prevent damage. These liquids carry away loose particles of abrasive which would otherwise clog the wheel or paper, or stick in the surface causing scratching. Liquids can also act as **lubricants**, lessening the effect of the abrasive particles. (You can demonstrate this by experimenting on scrap pieces of metal with the same abrasive used wet and dry.) Remember though that water, solvents or oils will have other effects on the object and that it may not be safe to leave them behind; they themselves may have to be cleaned off.

lubricants

Abrasion by rubbers (erasers) or rubber powder in cleaning paper objects is rather a different process which is used to remove relatively loosely held dirt. The stickiness of the rubber picks up and holds the dislodged particles, preventing them sticking back onto the surface again. Some erasers contain an additional abrasive, to make them more effective in removing ink, but these are usually too rough for conservation use, where original ink and pigment are to be left untouched.

A4 Vibrations

Another way of dislodging unwanted particles of dirt or corrosion is to use a vibrating tool to break up a dirt layer. Electrical pulses are used in some commercially available tools to produce vibrations in points (needles etc) of different sizes for cleaning and engraving. These tools are essentially automatic versions of the hand-held needle used for picking. It is necessary to be careful that the object does not itself get vibrated so much that damage is caused and to stop at cleaning and not start engraving.

ultrasonic cleaning

Ultrasonic cleaning also uses vibration. The vibration is produced electrically in specially made crystals (in the transducer) and is transmitted through metal plates to the liquid in the cleaning bath. This is usually water but organic solvents may be used. Sound waves are carried by the alternate compression and expansion of the liquid. If this alternation is rapid enough, the intense waves of vibration travelling through the liquid tear holes in it. Cavities of vapour appear and collapse at ultrasonic frequency – a phenomenon known as **cavitation**. These "ultrasonic bubbles" have a brushing action on the surface of an object placed in the tank because as the bubbles collapse, the liquid, locally, moves very fast. The process can be thought of as abrasion with molecule-sized grit.

cavitation

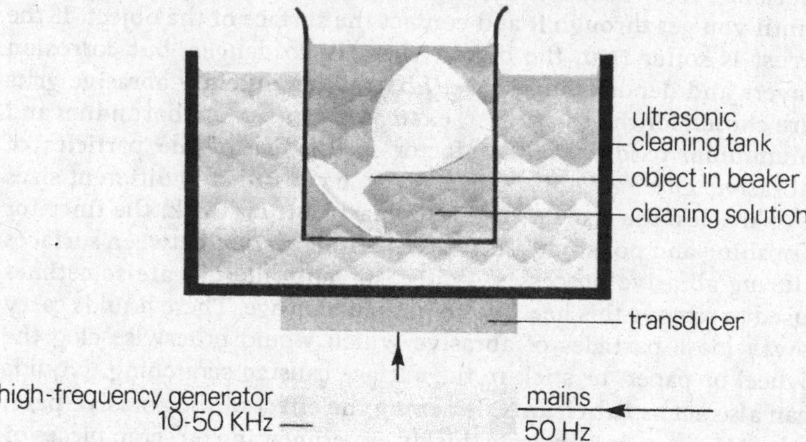

Figure 2.3 *Diagram of an ultrasonic tank.*

Ultrasonic equipment can be used to clean otherwise inaccessible places, for example inside narrow-mouthed vessels. The method has been used by conservators on metal objects and on stained glass but should not be used where there is loose, flaking decoration which might also become detached with the dirt. It will not remove dirt which is anchored to the surface with grease unless solvent or detergent is used in the bath. Ultrasound is not effective for cleaning lead, because the metal is so soft and malleable that it "soaks up" the vibrations. This indicates again, as with other mechanical cleaning treatments, that the physical properties of the original object are as important as those of the dirt layer. The equipment used by dentists to descale teeth is also used in conservation. This has an ultrasonic vibrating head immersed in a spray of water which flows round and through it. The vibration is transmitted into the water layer creating movement, vibration and cavitation, and thus cleaning the surface.

B Energy: familiar word – new concept

Book 1 concentrated mainly on descriptions of what matter *is*. The next stage, which is even more useful, is to learn to discuss scientifically how things *change*. Scientists communicate their ideas about change, or stability – which is the lack of change – by using the concept of energy. You need to learn what scientists mean by energy and what their underlying assumptions about it are, just as in Book 1 you learnt how they describe what things *are* in atomic terms.

There is an everyday usage of the word energy, as in "an energetic person" which contains the essence of the scientific meaning, for such a person "gets things done". Scientifically the idea is that it is *exchanges* of energy between matter which cause change. Incidentally, the study of the relationship between energy – especially heat – and matter is called **thermodynamics**. You know that by buying various commodities – coal, petrol, electricity – you are effectively buying energy to make use of it. However, you need a device which, by processing the fuel, "gets things done". And you know that energy is required to create change; to make things move or to make a liquid boil. A job such as boiling water can be accomplished in different ways – using gas, electricity or a wood fire. Moving a load can be done by people, horses, steam engines or diesel lorries. The same job requires the same amount of energy, but there are different ways of getting it done. *Energy* expresses the possibilities of what can be done. Seen in this way as something which is needed, supplied, converted or used in a series of events, energy is not entirely unlike that familiar currency – money. To help you to see how the rules for using the energy idea make sense, it may help to draw an analogy between transactions with money and changes in energy.

thermodynamics

Money flow in a small business

Imagine you are running a small business, such as a commercial conservation workshop:

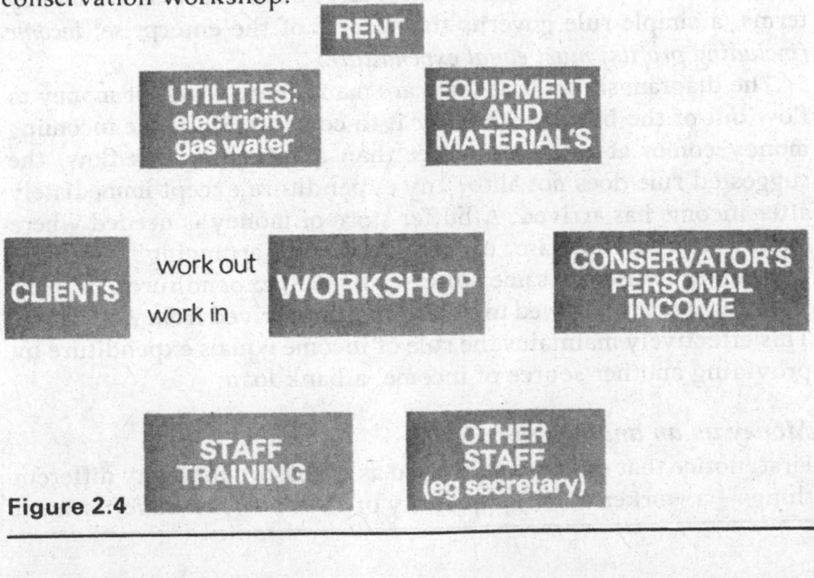

RENT

UTILITIES: electricity gas water

EQUIPMENT AND MATERIAL'S

CLIENTS

work out

work in

WORKSHOP

CONSERVATOR'S PERSONAL INCOME

STAFF TRAINING

OTHER STAFF (eg secretary)

Figure 2.4

The diagram shows the flow of goods, services and skills in and out of the workshop. The significant thing here is that there is no way of comparing one item with another. To make observations or predictions about the success of the business the different factors must be measured, and measured in units that can be compared. A litre of solvent, an hour of work, a cubic foot of gas, cannot be compared but their cost can. If the flows of materials, time and services are costed then the consequences of changing the input of one of them can be estimated. A second diagram can be drawn which includes this factor:

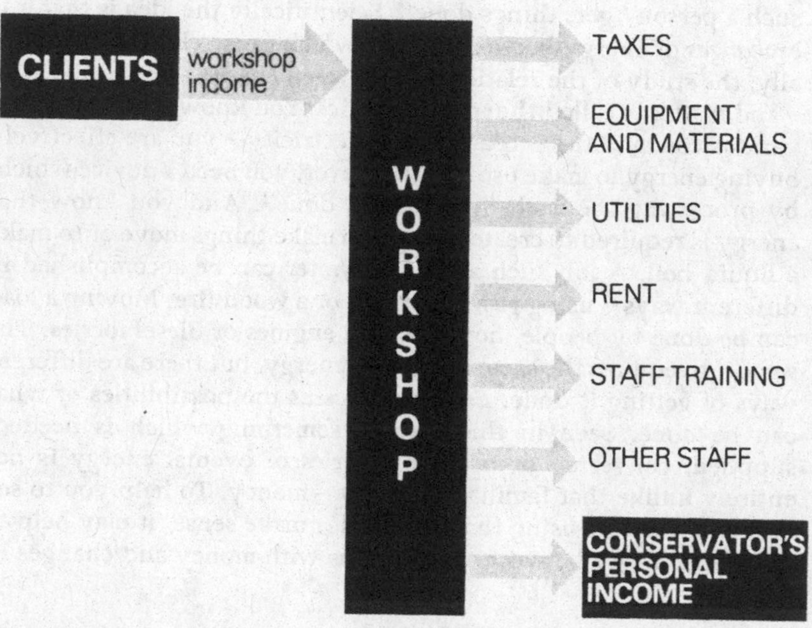

Figure 2.5

Once all the facets of the business have been expressed in money terms, a simple rule governs the success of the enterprise: *income (including profits) must equal expenditure*.

The diagram shows that there are many more ways for money to flow out of the business than for it to come in. Since the incoming money comes at intervals rather than as a continuous flow, the suggested rule does not allow any expenditure except immediately after income has arrived. A buffer store of money is needed where money can accumulate to be spent. By arranging for many businesses to use the same store – a bank – expenditure can, temporarily, even be allowed to exceed income derived from work done. This effectively maintains the rule of income equals expenditure by providing another source of income, a bank loan.

Money as an analogy to energy

First, notice that money can be used as a measure for many different things – a worker's time, a quantity of chemicals, an advertisement in a journal or being connected to the water supply. As transactions are made, *what the money represents* changes its form. Energy too has many forms; some chemical fuels and electricity have been mentioned already; heat and light are further examples. Matter in orderly motion, such as the wind or a flowing stream is a form of energy as is matter in a state of stress – a compressed spring or a stretched rubber band. Just as economists see the interdependence of human activities as represented by flows of money, so scientists see events in the physical world as flows of energy. Consider again the propagation of a crack as a chip of crusty dirt is pinged away:

forms of energy

Stage 1 Chemical energy in your body is converted into the force you exert on a scalpel blade which moves into the encrustation.

Stage 2 As the blade is pushed down, a notch is formed. Material at the root of the notch is stretched – atomic bonds acting like little springs have energy put into them. Material at the sides of the blade is compressed – springs in another condition.

Stage 3 The crack runs on, two fresh surfaces are created – stretching energy is converted into surface energy (because the atoms at the surface do not have all their bonding requirements satisfied).

Stage 4 When the crack is complete, the compression energy becomes energy of motion as the compressed material makes the chip spring away from the blade.

Several kinds of energy are brought into this description. Energy is said to be "converted" from one form to another rather as cash from the bank is "converted" to a litre of solvent, which is "converted" into a completed job. Notice that at stage 2 above some of the material was compressed and some stretched (Figure 2.6): both being

energy conversion

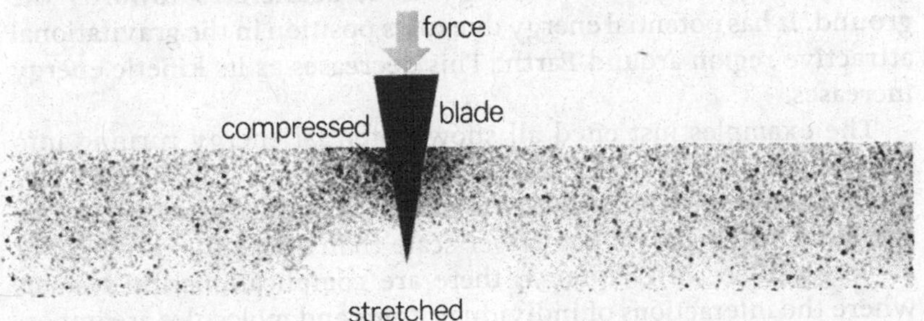

Figure 2.6 *Compressing and stretching forces caused by a blade being pushed down into material*

represented as a matter with energy put into it. The stretching energy was immediately converted but the compression energy was held until later, when the crack was complete, when it made the chip fly off. Here is the energy equivalent of the bank – a way of storing energy pending a later event. A fuel is even more like the bank, for what is stored can be spent in many ways at chosen times by anyone who can get access to it – for example via a cheque book for money from the bank, or via an engine for energy from a fuel.

The rule for spending energy is the same as that for money:, income must equal expenditure. In other words, if you add up all the energy put into an event and all the energy present at the end of the event the sums must be equal. You may be amused to learn that this rule is known as the **Principle of the Conservation of Energy**, and this is the first law of thermodynamics. A discussion of the units that are used to measure and compare different types of energy (a quantitive treatment) is not necessary here, but it is useful at this stage to make qualitative distinctions between various forms of energy. You saw that when a chip of encrusted dirt is removed, eventually the chip carries away energy because of its *movement*. This is known as **kinetic energy**. The amount of kinetic energy possessed by a body in motion depends on its mass as well as upon its speed. The other forms of energy in the example are not connected with motion. The energy held by stretched or compressed material is due, instead, to the relative *positions* of particles of matter. All such kinds of energy are known as **potential energy**. Potential, (capable of coming into action), implies that motion *could* result from this energy *if* a suitable mechanism existed. Thus the potential energy due to compression of material at the sides of a scalpel notch could only be converted into kinetic energy when the crack had extended all round the chip. Similarly a fuel has chemical potential energy – only convertible to kinetic energy by a mechanism such as an engine.

Principle of the Conservation of Energy

kinetic energy

potential energy

Another example you have met is potential energy due to the juxtaposition of two electric charges. If the bodies carrying the charges can move, the potential energy will be converted into kinetic energy – thus dust moves on to showcases. Gravity provides the most familiar example of potential energy. An object falling under gravity is gaining kinetic energy as it accelerates towards the ground. It has potential energy due to its position in the gravitational attractive region around Earth. This decreases as its kinetic energy increases.

The examples just cited all show potential energy turning into kinetic energy as soon as it is able to. It begins to look as if a tendency towards lower potential energy is what makes things happen. In the world of matter on a much larger scale than atoms this is true, but as you will see in Chapter 3, there are complications for systems where the interactions of individual atoms and molecules are important, as, for example, in chemical reactions.

The beginning of these complications comes when you think what happens so often to kinetic energy. Moving things generally slow down, because of **friction** (resistance to the movement), and this friction generates heat; to express this energy conversion you would say "kinetic energy becomes heat energy via the action of friction". However, as you will remember from Book I (Chapter 2), the basic

friction

nature of what we call **heat** must be described at the level of atoms and molecules, for heat is the kinetic energy of their random movement. You will remember that atoms and molecules are described as being in constant agitation, with collisions among them being the origin of chemical change. There is thus another tendency, which is for the **orderly motion** of larger things to become the **random motion** of atoms.

heat

orderly and random motion

Matter may possess energy either by virtue of *motion* or *position*, each of which may be *orderly* or *random*. These distinctions have a great deal to do with whether the matter's energy can be converted from one form to another. Observation has shown that there is a limit to the proportion of random energy (heat) which can be converted into any orderly form of energy. The **Second Law of Thermodynamics** expresses this limitation. The money equivalent is taxation. You would generally prefer the whole income of your workshop to be entirely your own wages. The conversion desired is

Second Law of Thermodynamics

completed work → personal income

expressed in the medium of money. It would be nice if the efficiency of the conversion could be 100%. But you know it cannot be. Figure 2.4 showed how many are the drains on money. Income has to be spent to achieve the job and some is even wasted, as far as the business is concerned, on taxes. Indeed, some of these unwanted expenditures (taxes, rates, interest) have priority claim on the workshop's income.

So with energy; when a machine is arranged to convert heat energy into another form, the conversion *cannot* be 100% efficient. There is an enormous "tax" on such a conversion. Even conversions where heat is not the source cannot be perfectly efficient. In cleaning the concern is that energy flows from the tool to the dirt and not into the object. Control of energy flows is therefore important. By and large control is easier if the *rate* of energy conversion is low, that is, if the process is gentle. The amount of energy that changes form every second is known as the **power** of the process and is measured in **watts**. This is a familiar measure of power; most electrical apparatus carries a label indicating its power in watts. For comparison a running man generates about 500 watts. When you see power ratings on a power tool in the range of several hundreds of watts and compare the delicacy of manual cleaning with the act of running, you should realise that using power-tools requires the control of a lot of energy per second. Where does it all go?

power

watts

Consider airbrasion as an example. The following diagram shows many of the routes for energy flow. Some of the observable results are what was intended, some are unimportant losses (wasted, but not harmful) but some must involve damage to the object. The point is that complex energy analysis can provide an assessment of a method without resort to trial and, perhaps, error, once you understand the process of conversion and the losses involved.

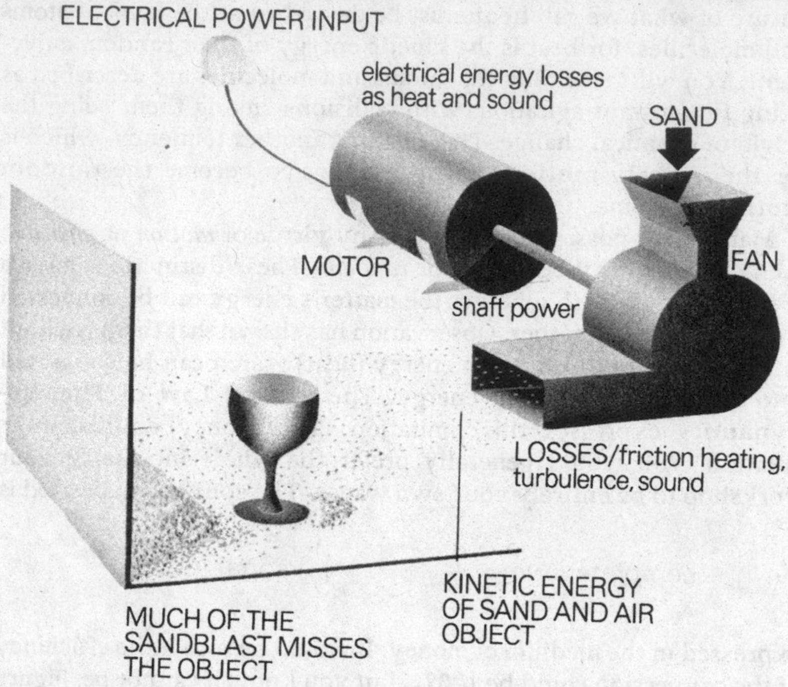

Figure 2.7 *A very simplified representation of airbrasive equipment, showing possible energy losses.*

The main virtue of the range of mechanical techniques for cleaning is that the task of separating dirt from object is closely under your control, as the operator, so that with skill, experience and care, you can make the energy flow where it is needed and make safe judgements about the amount of power to use.

3

Liquids and solutions

A Properties of liquids

A1 *Viscosity*

A2 *Capillarity and surface tension*

A3 *Volatility and drying*

B Describing solutions

B1 *What a solution is*

B2 *Measuring concentration*

B3 *Solubility*

C The science of solutions

C1 *Making a solution*

C2 *Energy changes in solutions*

Liquids and solutions

This chapter considers the nature of liquids and solutions, describing how and why they are formed and showing how this depends upon the secondary bonds which exist between molecules. From this, you will start to see which liquids are likely to dissolve which substances. The concept of energy, introduced in the last chapter, is expanded to explain its role in the formation of solutions.

A Properties of liquids

What are the properties of a liquid which are important in determining whether it is suitable for a particular cleaning job? Here are some, although you may be able to think of others.

- Will the liquid dissolve the dirt?
- Will it damage the object by reacting with it?
- Is the liquid runny enough to penetrate as required or sticky enough to stay where it is needed?
- Is the liquid volatile enough to dry off at a reasonable temperature or is it so volatile that the dirt will be redeposited during cleaning?
- Are the fumes from the liquid toxic and/or flammable?

The first two questions are obviously very important but to answer *all* these questions you need to know something about the properties common to liquids. A substance is recognised as being liquid by its ability to *flow* (this was discussed briefly in Book 1 in Chapters 2 and 4). However, a definition of the liquid state also needs some comment on the formation of a *surface*, since having a recognisable surface is what distinguishes liquids from gases, which also flow. The chemical properties which will tell you which liquid will dissolve which substance, and the physical properties, such as *viscosity* and *volatility*, can be qualitatively understood in terms of secondary bonding forces between molecules.

A1 Viscosity

viscosity

A liquid that is relatively immobile (not "runny") is said to be *viscous*. **Viscosity** is the scientific measure of a liquid's mobility and is an important property to consider when using liquids for conservation treatments.

To explain the flow properties of liquids it is necessary to visualise liquids at a molecular level. Flow implies that molecules are able to move past each other easily. Molecules which are attracted to one another by the stronger secondary bonds will form viscous liquids. The hydrogen bonding between water molecules means that water flows less easily than, say, ether, acetone or hexane which, although their molecules are bigger and heavier, possess only weak secondary bonds between their molecules. Concentrated acids (such as glacial acetic acid) are notably viscous because of their hydrogen bonding. Liquids containing long molecules which get entangled with one another also have high viscosities. This is why lubricating oils (with chains of 16–20 carbon atoms) are more viscous than their closely related hydrocarbon solvents (such as hexane which is only six carbon atoms long).

Sometimes it is necessary to control viscosity. If you want to increase the viscosity of a liquid, something must be dissolved in it which may use either, or both, of the properties mentioned above. Methoxy cellulose which has long molecules *and* the potential for hydrogen bonding is a common additive used to make a solvent less mobile. For example, it is added to the very mobile active ingredient, dichloromethane, CH_2Cl_2, to make paint stripper.

Temperature also affects viscosity. Liquids become less viscous as the temperature rises; animal size is an example where this is quite marked. A great change in viscosity occurs when a solid melts and turns into a liquid. The same forces of attraction which held the molecules in place in the solid are still there in the liquid, because in changing from solid to liquid state there are relatively slight changes in the separation of the molecules. However, if you were to visualise the behaviour of one particular molecule in a liquid, you would see it changing its loosely bonded association from one neighbour to another and then another. This contrasts with the solid state in which atoms or molecules only rarely change their relative positions (save for their constant vibration). The effect of increased molecular motion produces lower viscosity.

A2 Capillarity and surface tension

capillarity

Liquids are drawn spontaneously into very fine tubes or pores. This "blotting paper" effect is known as **capillarity** and is important in conservation in many ways. For example, it is used in blotting up liquids such as water or molten wax. In cleaning it transports solvents to inaccessible dirt in fine crevices or pores and carries dirty solvent off surfaces to which poultices have been applied. It can also act destructively; if it carries water into cracks in objects it may help to cause corrosion or, if it freezes, the water will expand and cause stresses which break the object.

To explain capillarity you have to think of the forces between the two different kinds of molecules – those in the liquid and those in the material of the tube or pore. There are attractive forces between *all* molecules; the important question is, which ones are the stronger – the *cohesive* forces between the molecules of liquid or the *adhesive* forces between the molecules in the liquid and those in the solid? The balance between these forces determines whether the liquids will be drawn up a fine tube and what the shape of the curved surface (called a **meniscus**) will be where the liquid touches a solid. There are two cases:

meniscus

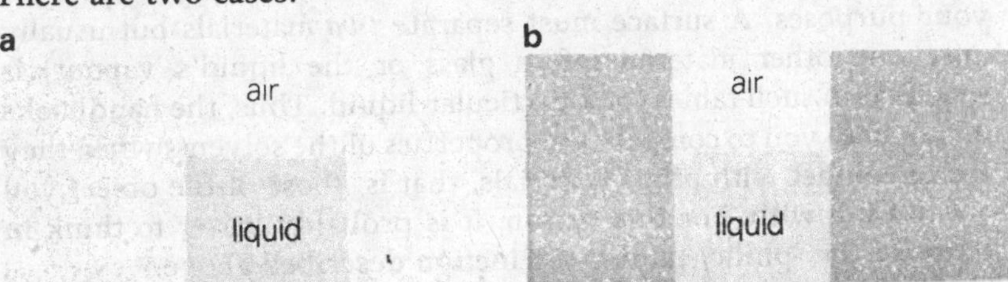

Figure 3.1 **a** *A concave and* **b** *a convex meniscus of a liquid.*

Where there is a strong attraction between the liquid and the solid, the liquid climbs up the solid's surface as in (a). Where the balance lies the other way, and the cohesive forces are strong but the adhesive forces weak, the liquid pulls in towards itself as in (b).

Strong capillarity is to be expected with polar liquids which, although they cohere strongly, may offer a mechanism for strong intermolecular forces from solid to liquid. If the solid can offer hydrogen bonds to a polar liquid (for example water) there will be *absorption*. Such materials (for example, cotton, wood and proteinaceous materials) are **hydrophilic** which means water-loving. Greasy surfaces repel water and are termed **hydrophobic** or water-hating. As they cannot make hydrogen bonds with grease the water molecules cohere together. Most natural waterproofing methods (such as those of furs and feathers) depend on coatings of grease; coatings of non-polar molecules like paints and varnishes have a similar waterproofing effect.

hydrophilic and hydrophobic materials

You will often have noticed droplets sitting on fibres, failing to wet them or penetrate. This condition, which obviously prevents cleaning, comes about because the forces of *cohesion* between the water molecules are stronger than the forces of *adhesion* between water molecules and fibre molecules. The role of soaps and detergents as additives to water is to ensure that the forces at the surfaces where solid and liquid meet are better matched (hence their general name of **surfactants**, meaning that they act on the surface).

surfactants

surface tension Descriptions of these effects are often given in terms of **surface tension** (or sometimes *surface energy*). Cohesive forces in a liquid have the effect of pulling its molecules as close to one another as possible. For a small quantity of liquid this tendency will produce a spherical drop, since in this shape every molecule is as close to all the others as it can be. This shape also has the smallest surface area for a given volume. Imagine the liquid surrounded by a stretched skin pulling it into shape, rather as if the liquid was in a balloon. There would be a certain stretching force or "tension" in the skin *equivalent* to the cohesive forces. A change of shape, say from a droplet standing on a surface to a flat puddle, will involve an increase in surface area and hence would require work (energy put in) to stretch the skin. It is easier for a droplet to spread and wet the surface if its molecular cohesive forces are weak *or* if the skin tension is low. These are two ways of saying the same thing.

Although data for the surface tension of water (and for other liquids) are given in handbooks, this is not often very helpful for your purposes. A surface must separate *two* materials but usually only one other material (often glass or the liquid's vapour) is specified in such tables for a particular liquid. Thus, the handbooks do not help you to compare the properties of the solvents when they are in contact with other materials, that is, those of the object you are dealing with. For this reason, it is probably better to think in terms of the -philic/-phobic distinction described above.

A3 Volatility and drying

After any liquid cleaning process it is usually important that the object is dried. A liquid which rapidly evaporates off a surface is said to be *volatile*. When a liquid evaporates its molecules become separated from one another and the liquid no longer coheres. Thus, to be *volatile* a liquid must have weak intermolecular forces. Volatile liquids usually have light molecules, and so the same amount of energy will make them move faster than heavier molecules or long

volatility tangly ones. As with viscosity and surface tension, **volatility** also relates to the way the molecules hang on to one another. Hence liquids which have low viscosity and surface tension (such as acetone) tend also to have high volatility and, vice versa, liquids with high viscosity and surface tension (such as water and glycerine) have low volatility.

In cleaning, if a solvent evaporates too quickly the dirt it has taken up will be redeposited on the object (perhaps making matters worse in the process if the solvent has carried the dirt into crevices). However, if after cleaning the drying process is too slow and heat is needed to speed it up, the object may again be put at risk. One useful way round this problem is to clean with one solvent, and then to wash away that solvent with another one which is more volatile. Using acetone to help dry something wetted with water is an example. An evaporating solvent such as water will tend to saturate the air around the object; using a fan to create an air flow can help to speed up the evaporation process.

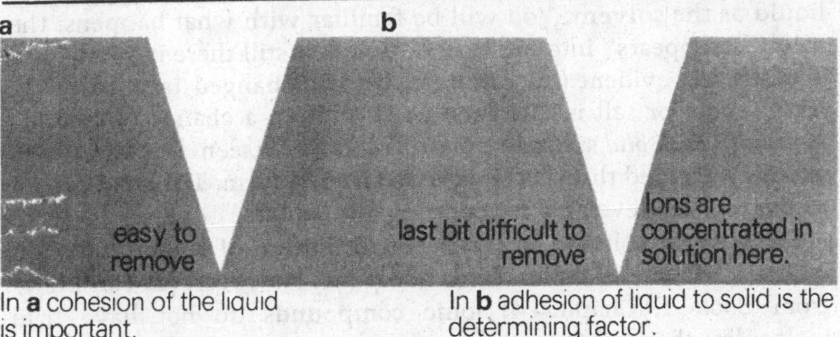

In **a** cohesion of the liquid is important.

In **b** adhesion of liquid to solid is the determining factor.

Figure 3.2 *Water trapped at the bottom of a corrosion pit.*

Water, as a relatively non-volatile cleaning fluid and as a polar solvent which dissolves ionic substances, presents its own problems. The least bit of water remaining in a corrosion layer on metals, especially if not all the salts have been removed, may be enough to start up the corrosion all over again. At the bottom of a corrosion pit the walls are close together and the layer of water very thin, so that in addition to the physical difficulties of getting the water out, the chemical bonds extend from the solid to each individual water molecule and prevent the water leaving. (This is the same adhesion between solid and liquid that occurs in surface tension.)

To avoid leaving salts behind as the water dries off, they must either be totally washed out or deposited elsewhere (for example in a poultice of paper pulp or a mud pack).

B Describing solutions

B1 What a solution is

Any mixture of substances which is homogenous down to a molecular scale qualifies for the name of **solution** though the word is commonly used for solids dissolved in liquids. Before concentrating on these, remind yourself of a few examples of non-liquid solutions. First, a mixture of gases, such as air (mainly nitrogen and oxygen) can legitimately be regarded as a solution because the molecules are intimately and randomly mixed. Air is more or less homogenous; you would not find pockets of pure oxygen and others of pure nitrogen even if a very small sample (a few million molecules) was analysed. Solid solutions include several metal alloys. Electrum (silver and gold) is one example, cupro-nickel (copper and nickel) is another. Gases can dissolve in liquids; the amount of gas dissolved can be increased by increasing the pressure. The fizz when a bottle of champagne is opened is carbon dioxide coming out of solution as the pressure is released. Liquids can also dissolve in liquids. If two liquids combine to form a soluton they are **miscible**. Ethyl alcohol and water are miscible but toluene and water are not.

solution

miscibility

The more familiar solutions of solids in liquids need to be described in more detail and it will be helpful to define some words before explaining what happens when solutions are formed. When a liquid dissolves a solid the solid is known as the **solute** and the

solute

solvent

liquid as the **solvent**. You will be familiar with what happens; the solid "disappears" into the liquid. That it is still there in some form is often self-evident (for example, by the changed taste of water when sugar or salt is dissolved in it or from a change of colour). Although the *solid* substance is no longer to be seen, it is still there, so finely divided that it no longer has a solid form. The solid can be recovered, however, by evaporating the liquid.

For a sugar solution in water the definition of a solution as an intimate mixture of molecules is adequate. However, (as you know from Book 1, Chapter 4) ionic compounds do not have true molecules: their solid crystals are organised arrays of positive and negative ions. When such a substance dissolves, the ions are insulated from one another by the water and so the mixture exists at a *sub*-molecular level. The evidence for the independence of ions in solution has already been mentioned in Chapter 2 where it was explained that ionic solutions are capable of conducting electricity. Of course, when the solvent is evaporated from an ionic solution, the ions come together again and, because of their electrical interaction they form crystals.

B2 Measuring concentration

There are two things needed to describe a mixture; what its *constituents* are and what the *proportions* of each constituent are. Unfortunately no single convention has been chosen to describe the

concentration

concentration – the proportions of each constituent in solutions. One way would be to state the relative proportions of molecules of each type in the mixture. However, chemists measure the concentration of the solution, by working out the number of moles of *solute* in one litre of *solution*.

The practical good sense of using molar concentrations (see Book 1) rather than molecular proportions lies in the way solutions are used. For liquid solutions, the obvious measure of quantity is volume. This could be measured roughly in spoonfuls, more accurately with a graduated cylinder and most accurately, when necessary, with a pipette or a burette.

When solutions are used as chemical reagents (chemically reactive substances), it is the solute which is doing the reacting; the solvent is merely a vehicle to carry the solid into the reaction in its ultimately finely divided form. Thus what you need to know is the number of *moles* (see Book 1, Chapter 3) of *solute* you are providing for the reaction. You will remember that given a chemical equation for the intended reaction, you can use one piece of information (how many moles of one reactant you have) to calculate another (the maximum amount of the other reactant which will be consumed or the amount of product formed).

Specifying solutions

In practice, however, the exact quantity of one of the reactants, the dirt, will not be known, nor will its exact chemical nature. It is pointless to measure the number of moles of solute for a cleaning reaction so you need not bother with the calculation of molecular mass and so forth. But you *do* have to make up solutions exactly and reproducibly, so there *must* be some convention for describing the concentration of solute. Clear instructions can be given by specifying the *weight* of solid to be present in a given *volume* of solution. This is the simplest and most easily understood measure. A convenient and universally accepted description is the number of grams of solute in a litre of solution, written as so many g/l (**grams/litre**). **grams/litre**

You must remember, however, that by convention concentrations are always described as so much solid per unit volume of *solution* and *not* per unit of *solvent*. This means that to make up a solution to an exact concentration you have first to dissolve the weighed amount of solid in a small amount of solvent and then dilute this solution to the prescribed volume.

The descriptive form, grams/litre, just explained, is unambiguous and therefore unlikely to cause mistakes. There are other less straightforward ways of describing solutions. Loosely to talk of a solution of "so many per cent" is particularly confusing. A "ten per cent solution" might refer to a *weight* per volume per cent solution or a *volume* per volume percentage, depending on the solute. A 10% solution of sodium hydroxide means that 10g of solid has been dissolved to make 100ml of solution (w/v%). A 10% solution of ethanol could contain 10ml of liquid alcohol diluted to 100ml of solution (v/v%). The w/v% convention can cause confusion as the units of measurement are not stated. A 10% solution of salt in water *could* be 10lbs of solid in 100 gallons, which gives a solution only one tenth as concentrated as 10g in 100ml. So for consistency the units should be stated, in which case you might just as well have stated the concentration in g/l.

B3 Solubility

There is a limit to the amount of solute which can be dissolved in a given amount of solvent. This maximum amount of solute is termed the **solubility** of the solute. It varies from solute to solute and from solvent to solvent. The solubility of solids *increases* with rising temperature. (For gases dissolved in liquids solubility is *less* at higher temperatures.) A solution which contains as much solute as possible is called a **saturated solution**. (The same word is used to describe air that can hold no more water vapour, because the water vapour is a solute in air.) **solubility** **saturated solution**

In cleaning applications, provided the dirt can dissolve in the solvent to some reasonable extent, the actual solubility does not much matter; extra solvent can always be made available. If you choose to use a warm solvent, it is more likely that you intend to increase the *speed* at which the dirt is dissolved rather than the total *amount* the solvent can hold. What *is* important, on the other hand, is the solubility of the *object* in the solvent. Strictly speaking there is no such thing as an **insoluble** substance. Every solid will dissolve in every liquid to some extent. It is vital that the solubility of any part of the object in the fluid chosen for solvent cleaning should be **insolubility** very, very small. Only then can you afford to immerse the object in large quantities of solvent. There are obvious examples where an object's solubility is too high, such as alabaster in water, limestone and bronze in acids. There are many other examples of solvents that attack both dirt and object. Many stains and dyes are both affected by the washing of textiles. Published recommendations to use unusual solvents, such as N-methyl-2-pyrrolidone for dissolving starch away from cellulose, usually stress the problems of safeguarding the object rather than those of attacking the unwanted material. Starch and cellulose are very similar chemically and this solvent has been chosen because it is more selective than any other. It may not be the worst solvent for cellulose nor the best for starch but it is one of few compounds which distinguishes between them.

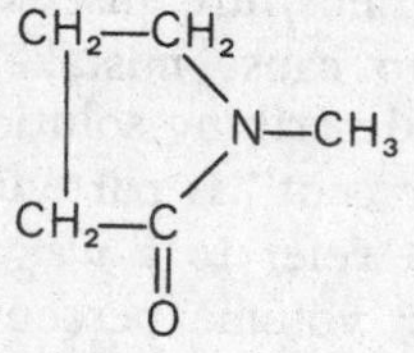

$$\begin{array}{l} CH_2{-}CH_2 \\ \quad\quad\quad\quad N{-}CH_3 \\ CH_2{-}C \\ \quad\quad\;\; \| \\ \quad\quad\;\; O \end{array}$$

Figure 3.3 *The structure of N-methyl-2-pyrrolidone. The* pyrrolid *part of the name refers to the five-membered ring containing a nitrogen atom. The* 2 *and the* -one *say that the second atom counting round the ring from the nitrogen has a double bonded oxygen attached. The* N-methyl *tells you there is a CH_3 group joined to the nitrogen atom.*

For cleaning, therefore, you must find a solvent which:

- does not dissolve to a significant extent any part of the object.
- dissolves the dirt or whatever is holding the dirt in place.
- dries out, but not so quickly that it leaves dirt infiltrated into the object.

C The science of solutions

C1 Making a solution

When a solid is put into a liquid the significant events occur at the interface between the two substances. You can think of what goes on using Figure 3.4:

The kinetic energy of the molecules of both liquid and solid results in some of the liquid molecules worming their way into the solid (stage 2). As this effect increases the solid swells because the molecules of the solid move apart. As the solvent penetrates further, molecules of the solid lose contact with their neighbours and join in the random wriggling motion of the liquid. In going from stage 1 to stage 2 a molecule of the liquid, which happens to be moving fast enough in the right direction to penetrate the solid's surface, has broken the strong bond between two "solid" molecules and

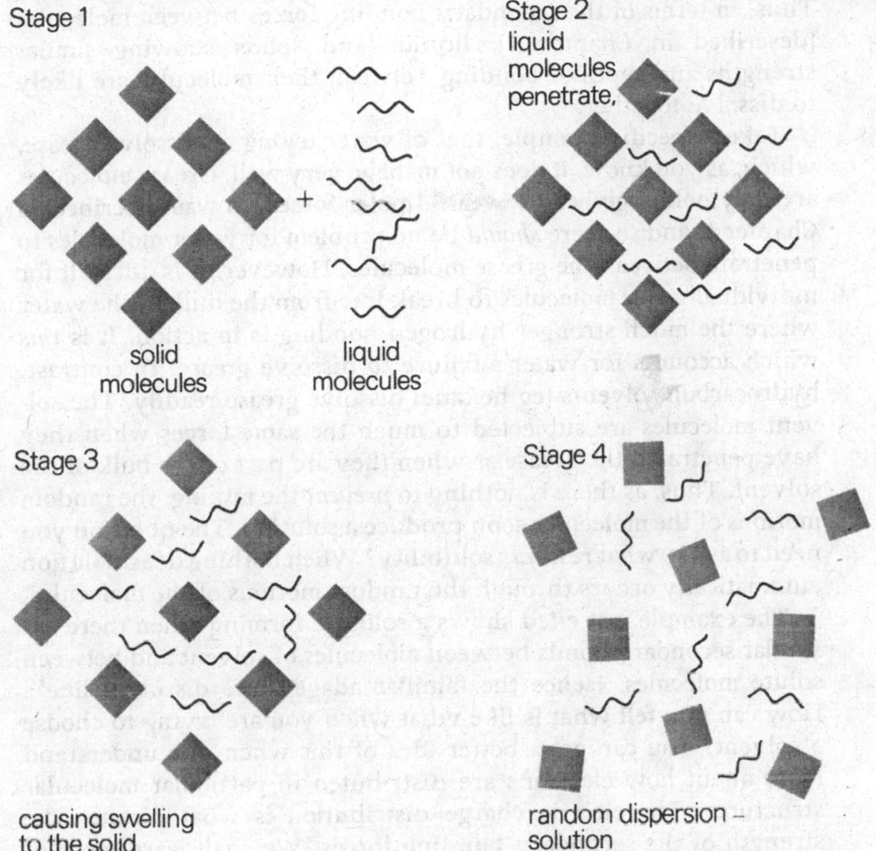

Figure 3.4

replaced it by bonds between "solid" molecules and "liquid" molecules.

There are, however, two factors which are likely to affect this process. The faster the "liquid" molecules move the more likely it is that bonds will be broken and that the form of energy involved will change from kinetic to potential. The new bonds formed between "solid" molecules and "liquid" molecules need to be as strong as the bonds which previously existed between the molecules of the solid (which they now replace) so that the penetrating "liquid" molecules do not have the tendency to separate out again. Bearing in mind these considerations, you will see that the solid is more likely to dissolve if:

- **The temperature is high** since this means more molecules are moving fast.
- **The forces between the molecules of the liquid and the molecules of the solid are of similar strength.** This then means that the solid will be more likely to accept "liquid" molecules into it and that the "liquid" molecules are not strongly pulled back themselves to remain as a separate liquid.

Thus, in terms of the secondary bonding forces between molecules (described in Chapter 1), liquids and solids showing similar strengths and types of bonding between their molecules are likely to dissolve together.

Take a specific example, that of water trying to dissolve grease, which, as you know, it does not manage very well. Grease molecules are only held together by weak dipolar forces, as was described in Chapter 1, and so there *should* be no problem for water molecules to penetrate between the grease molecules. However, it *is* difficult for individual water molecules to break free from the bulk of the water where the much stronger hydrogen bonding is in action. It is *this* which accounts for water's failure to dissolve grease. In contrast, hydrocarbon solvents (eg hexane) dissolve grease readily. The solvent molecules are subjected to much the same forces when they have penetrated the grease as when they are part of the bulk of the solvent. Thus, as there is nothing to *prevent* the mixing, the random motions of the molecules soon produce a solution. The question you need to ask is; what *restricts* solubility? When nothing does, solution automatically occurs through the random motions of the molecules.

The example just cited shows a solution forming when there are similar secondary bonds between molecules of solvent and between solute molecules. Hence the familiar adage "like dissolves like". How can you tell what is like what when you are trying to choose a solvent? You can get a better idea of this when you understand more about how electrons are distributed in particular molecular structures. The electric charge distribution is what dictates the strength of the secondary bonding forces. We shall learn more of that in Chapter 4.

C2 Energy changes in solutions

You may have seen an article on choosing solvents in the journal *The Conservator**. It discusses the changes in energy that occur when something is dissolved. The key idea is expressed by an equation:

$$\Delta G_{mix} = \Delta H_{mix} - T\Delta S_{mix}$$

Like a chemical equation, this is merely a shorthand notation for something that could be expressed in words as a fairly lengthy sentence. Once the meanings of the symbols are known the equation can be used to predict behaviour or, more usually, to support descriptions of observed behaviour.

The triangle Δ is a capital delta from the Greek alphabet. It means "The change in . . .".

The subscript "mix" means "on mixing". This distinguishes the quantities in the equation from other similar ones relating to other changes such as chemical reaction or going from liquid to vapour.

* *Solubility Parameters and Varnish Removal: A Survey, G. Hedley,* The Conservator, *No. 4, 1980, p. 12.*

G is the symbol for something called *free energy*.
H is the symbol for something called *enthalpy*.
S is the symbol for something called *entropy*.
These terms will be explained later.

T is the temperature.

$T\Delta S$ means T multiplied by ΔS.

So ΔG_{mix} is the symbol for "The change in **free energy** on mixing" (say, on mixing water and salt, or resin and toluene). The equation suggests that ΔG_{mix} can be calculated by a simple subtraction. If numerical values for the changes in enthalpy and entropy on mixing can be found, the equation can be used to determine a value for the free energy change.

free energy

Although the words and symbols may all be unfamiliar the mathematics is no more complicated than saying: "The money I hand over to the shop assistant *minus* the cost of the bought article *equals* the money the shop assistant hands back to me." I will get some money or no money back depending on the amount I paid and the amount I needed to pay. Indeed I may be asked for more.

If $T\Delta S_{mix}$ is numerically equal to ΔH_{mix} there will be nothing left after the subtraction. ΔG_{mix} will be zero. But if $T\Delta S_{mix}$ is smaller than ΔH_{mix} there will be something left over, so ΔG_{mix} will be a positive amount. If, on the other hand $T\Delta S_{mix}$ is numerically greater than ΔH_{mix} the result of the calculation

$$\Delta H_{mix} - T\Delta S_{mix}$$

will be a negative value for ΔG_{mix}. The reason for these calculations is that if the value for ΔG_{mix} is found to be negative, a solution *should* be formed. If the result is zero or a positive number then the mixed substances will *never* form a solution.

This result is directly related to the **changes of energy** in the mechanical systems we discussed in Chapter 2. As a compressed spring is released or as a brick falls its potential energy (PE) decreases. The change in potential energy, ΔPE, on going from an initial state to a second state is negative. If ΔPE is negative the event described is one that happens spontaneously. Events for which ΔPE is positive (springs compressing themselves, bricks falling upwards) do not occur readily.

changes of energy

Because of the great thermal agitation of molecules it is possible that individual particles may be knocked from a state of low potential energy to one of high potential. However, the general trend of the energy change in the bulk material will be in the reverse direction. The quantities ΔG, ΔH and ΔS refer to the overall state of billions of particles and not to individual changes. At the molecular level there are two conflicting tendencies:

- **those preventing solution** – the tendency of the attractive forces to hold together the molecules of the substance to be dissolved.

- **those encouraging solution** – the tendency of molecular motion to mix things up.

If the equation is to predict that solution will take place when the answer is negative, then the mixing up tendency must have a negative sign in front of it and the holding together tendency a positive sign, so that if the mixing tendency is the greater the answer will be negative.

H: The enthalpy

enthalpy The **enthalpy** (H in the equation) is the *total* energy of all the particles in the system. When two substances are mixed and then form a solution there is no increase in the total number of particles. If the temperature remains the same the total kinetic energy of the particles is not altered. So the only change is in the inter-particle potential energy. If the molecules or ions in a solid are strongly bonded then energy is required to break them apart. There is an increase in potential energy.

Since the breaking of bonds inhibits formation of a solution the change in enthalpy is given a positive sign in the equation. Changes in enthalpy are sometimes detected as heat absorbed or evolved on forming a solution. A positive enthalpy change is shown by the solution getting colder.

S: The entropy

It is more difficult to express in terms of energy the tendency of molecular motion to mix the molecules. That this tendency depends upon temperature through the agency of molecular kinetic energy cannot be doubted. But temperature is *not* energy and does not

entropy measure the amount of mixing produced by dissolving. **Entropy** (S in the equation) is the quantity which fulfils the connecting role. Suppose, for instance, sugar is dissolved in water:

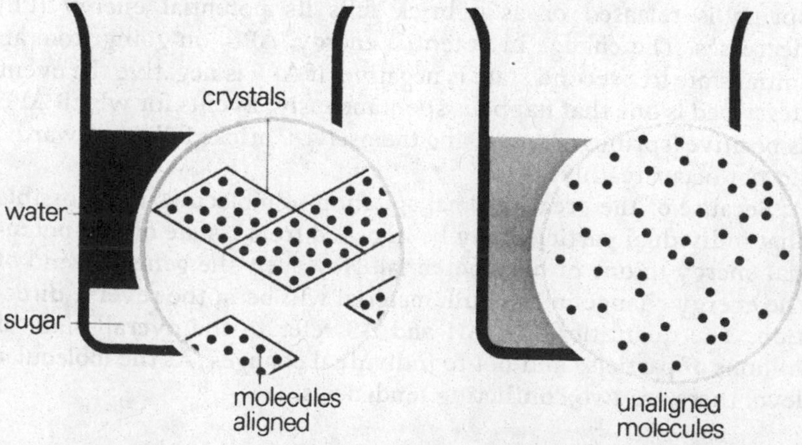

Figure 3.5 *The rearrangement of molecules in a solid to form a random spread after dissolution in a liquid. The insets show diagrammatically magnified portions of the two states.*

Before the sugar dissolves, its molecules are in regular crystal patterns and quite separate from the water. *After* dissolving the molecules are randomly spread through the water. Many ways could be invented for describing the order of "before" and the disorder of "after". For example the length and angles between lines joining some number of molecules could be averaged (Figure 3.6):

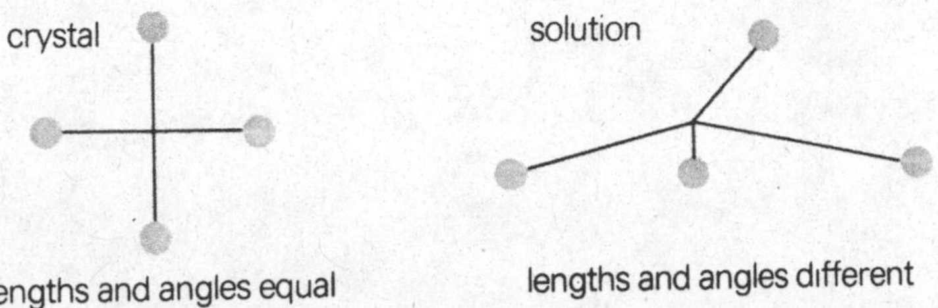

Figure 3.6 *Hypothetical means of describing the order and disorder of solid and dissolved crystals.*

Entropy is one possible description of the extent of disorder in a system. The mathematics that relates the number of possible states of a system such as a solution and compares the order in a crystal with the disorder in solution is very complex. What you need to remember is:

- Entropy measures disorder.
- Entropy is greater for disordered conditions than for orderly ones.
- The usual tendency for mixing by molecular motions makes entropy increase. Another way of stating the Second Law of Thermodynamics (Chapter 2) is to say that there is a general tendency for entropy to increase.

With this information the free energy equation can, at last, make sense.

$$\Delta G_{mix} = \Delta H_{mix} - T\Delta S_{mix}$$

will come out negative if a big change in entropy, ΔS_{mix}, can coincide with a small change in molecular potential energy, ΔH_{mix}. Mixing a given number of molecules in a given volume will produce more or less the same ΔS_{mix} whatever the molecules may be. So if they happen to be ones which cohere tightly (ΔH_{mix} large) you may have to increase the temperature T to make the negative part of the equation swamp the positive part – or, in reality as opposed to in the *model* of reality, you may have to increase the temperature to make a solution.

All these general ideas about liquids and solutions eventually come together when a choice is made of a particular solvent for a particular job. The following chapters look at some of the options which are available to you.

4

Organic solvents

A General observations

B Classes of organic solvents

B1 *Non-polar hydrocarbon solvents*

B2 *Halogen-substituted hydrocarbons*

B3 *Polar solvents containing oxygen*

Organic solvents

This short chapter reviews the structures and properties of some organic substances which are liquid at ordinary temperatures and are used as solvents for cleaning. The division into *polar* and *non-polar* categories allows you to apply the idea that like substance will dissolve like (introduced in the previous chapter) with a little more precision. It also enables you to explore further the characteristics of groups in organic molecules.

A General observations

If like substances dissolve like, then you will expect to use organic solvents on organic dirt (such as grease, old varnish, paints and glues, tar and mildew stains). Since this type of dirt is often found on objects made of organic materials, it may be a tricky matter to choose a solvent which is both effective and safe. However, substances from which such artefacts are made tend to have giant molecules with either primary or strong secondary bonds linking them. These are more difficult to dissolve than the small dirt molecules. The consequence for cleaning is that organic solvents, in general, are likely to be innocuous as far as any risk of dissolving the main structural materials, such as proteins or cellulose, are concerned. Nature's solution is to use enzymes (see Chapter 7). You can be rather less sure, however, about the supplementary materials of an object, dyestuffs, varnishes, essential oils, and so on. The solvents can be categorised by the type of secondary bonds they can make. If you know solvent A is unsafe in such and such a circumstance you won't ignorantly try solvent B which is similar. Or, if you know A works but is a bit too volatile, you will know which substances are similar in solvent action but are less volatile.

You will remember from Chapter 1 that the main subdivision of secondary bond types is between *non-polar* molecules which show *only* the weak Van der Waals attraction, and *polar* molecules which have additional stronger electrostatic attractions due to their uneven electric charge distributions. Since we have seen that physical properties as well as chemical ones are under the control of these bonds it will be valuable first to see the effects at work on a simply measured property – the boiling point (which at least partly is a comment on volatility). Weak bonds and light weight should make for low boiling points: opposite tendencies give high-boiling liquids. So if you look at families of compounds with similar bonds the boiling point should go up with molecular mass. Here are two sets of examples:

Hydrocarbons (paraffins) – non-polar molecules

Name	Formula	Mol. mass	Boiling point °C
methane	CH_4	16	−161
ethane	C_2H_6	30	− 89
propane	$CH_3CH_2CH_3$	44	− 42
n-butane	$CH_3CH_2CH_2CH_3$	58	0
n-pentane	$CH_3CH_2CH_2CH_2CH_3$	72	+ 36
n-hexane	$CH_3(CH_2)_4CH_3$	86	+ 68

In fact all these are too volatile to be considered as solvents but the last of these is the lightest to stay liquid long enough to dissolve anything. Compare those boiling points with those of a family of polar molecules, noting the similarities in molecular mass.

Alcohols – polar molecules with H-bonding

Name	Formula	Mol. mass	Boiling point °C
methanol	$CH_3{-}OH$	32	66
ethanol	$C_2H_5{-}OH$	44	78
1-propanol	$C_3H_7{-}OH$	60	97
1-butanol	$C_4H_9{-}OH$	74	118

It is clear that, weight for weight, the alcohols are far less volatile than the paraffins. We can ascribe this to the stronger cohesion between their molecules. Water (mol. mass 18) which also has an −OH group has an amazingly high boiling point (100°C) but of course *both* ends of its molecules can make secondary bonds to other molecules.

It is also instructive to compare boiling points of substances with similar molecular masses but different polar groups. Here are some examples, all with a molecular mass around 60.

Name	Formula	Mol. mass	Boiling point °C
butane	$CH_3CH_2CH_2CH_3$	58	0
methyl ethyl ether	$CH_3CH_2OCH_3$	59	11
acetone	$CH_3C(=O)CH_3$	58	56
1-propanol	$CH_3CH_2CH_2OH$	60	97
acetic acid	$CH_3C(=O)-OH$	60	116

These five substances have nearly the same molecular mass yet there is a range of boiling points of more than 100°C. This demonstrates the polarity associated with the electronegative oxygen atom bond in *ethers* and more especially when it is *double bonded* to carbon as in the *ketones*. The ability to hydrogen bond, provided by the – OH group, gives a higher boiling point, while the acid with both – OH and a strongly polar C = O double bond has the highest boiling point of all. Since volatility can be controlled both by molecular weight and strength of secondary bonding, subtle choices can be made within each class of solvent.

B Classes of organic solvents

B1 Non-polar hydrocarbon solvents

non-polar hydrocarbo

Because hydrocarbons interact only by Van der Waals forces they are liquid only when their molecules are quite heavy. Hexane is the lightest of the *straight chain* hydrocarbons which is sufficiently non-volatile to be of use in cleaning. Cyclohexane, C_6H_{12}, has molecules in the form of a *ring* (cyclo-, from Greek, a ring):

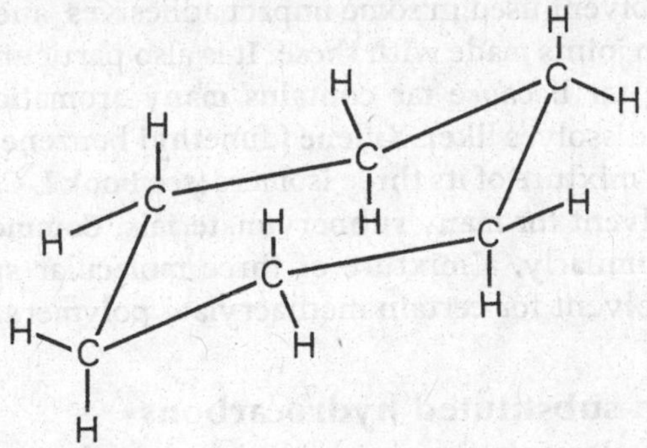

Figure 4.1 *The structure of cyclohexane.*

Because the bonds from the carbon atoms point to the corners of an imaginary tetrahedron (see Book I, Chapter 3) the ring is not flat. This is in contrast to the aromatic hydrocarbons (such as benzene, C_6H_6) where the six carbon atoms in the ring are all in the same plane. The molecular orbitals which connect all six carbons at once in aromatic compounds can only be formed if the molecule is flat.

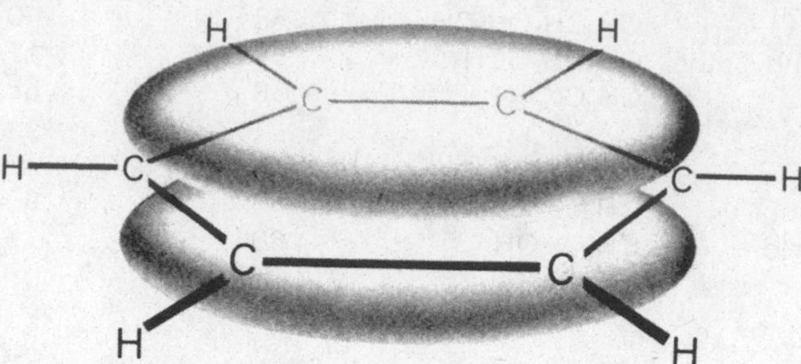

Figure 4.2 *Structure of benzene. Molecular orbitals are seen as clouds of electrons above and below the plane of the carbon and hydrogen atoms.*

The principal disadvantage of hydrocarbon solvents is that they are flammable, although they do find some uses in cleaning. They are useful for softening or dissolving greasy deposits which are, largely, hydrocarbons themselves (like dissolves like).

Pure hydrocarbon compounds may be too expensive for all but special applications but many cheaper mixtures are available. White spirit, solvent naptha, Stoddard solvent, etc., are mixtures of **aliphatic (alkyl)** and **aromatic (aryl)** hydrocarbons distilled from petroleum. The standard trade descriptions of these define only the flash point (the lowest temperature required to form a vapour/air mixture which will explode) and a range of boiling points. Within this range fall many different hydrocarbons so that the solvent strength may vary from batch to batch. Mixtures with more closely controlled composition, and hence more consistent solvent strength (such as Shellsol T), are made and sold by the oil companies.

liphatic and aromatic hydrocarbons

Of the aromatic hydrocarbons, benzene must be avoided in workshops because its vapour is very poisonous. Toluene (which is less toxic) is the solvent used in some impact adhesives, and so it may be used to loosen joints made with these. It is also particularly effective for removing tar because tar contains many aromatic compounds (another like dissolves like). Xylene (dimethyl benzene) is sold commercially as a mixture of its three isomers (see Book I, Chapter 5) and is a useful solvent for many rubbery materials. Commercial diethyl benzene is, similarly, a mixture of three molecular species and is used as the solvent for certain methacrylate polymers.

B2 Halogen-substituted hydrocarbons

The group of elements fluorine, chlorine, bromine and iodine are known as the **halogens**. If a hydrogen atom in a hydrocarbon is replaced by a halogen atom the compound is considerably less flammable (but these compounds are toxic). The series based on methane, CH_4, is:

halogens

CH_3Cl chloromethane (methyl chloride), a gas at room temperature.

CH_2Cl_2 dichloromethane (methylene chloride) used in commercial paint strippers (eg Nitromors).

$CHCl_3$ trichloromethane (chloroform) which you will recognise as an anaesthetic, as are most of these compounds.

CCl_4 tetrachloromethane (carbon tetrachloride).

There are many examples of halogen-substituted hydrocarbons based on ethane, containing two carbon atoms. Figure 4.3(a) shows 1, 1, 1 trichloroethane; the numbers 1, 1, 1 indicate that three chlorine atoms are on one carbon atom. This compound is used as a metal degreasing agent. Figure 4.3(b) shows trichloroethylene, one of the dry cleaning fluids.

```
    Cl  H                 Cl  H
    |   |                 |   |
Cl—C—C—H                 C=C
    |   |                 |   |
    Cl  H                 Cl  Cl
```

a 1, 1, 1 trichloroethane *"Genklene"* **b** trichloroethylene

```
    Cl  Cl               Cl  Cl
    |   |                |   |
Cl—C—C—F              F—C—C—F
    |   |                |   |
    Cl  F                Cl  F
```

c 1, 1 difluoro tetrachloroethane *"Vaclene"* **d** 1, 1, 2 trifluoro trichloroethane *"Arklon"*

Figure 4.3 *Some examples of halogenated solvents derived from ethane (two carbon atoms).*

Hydrocarbons containing chlorine and fluorine are much more volatile than unsubstituted hydrocarbons of similar molecular mass. Compare Arklon, $C_2Cl_3F_3$ (mol. mass = 188) shown in Figure 4.3 (d) which boils at 47°C with tridecane, $C_{13}H_{28}$ (mol. mass = 184), which does not boil till 243°C. The forces of attraction between the molecules in chlorinated and fluorinated hydrocarbons must be very weak. This may be because the carbon atoms are more or less completely surrounded by electron-grabbing halogen atoms, so the outer surfaces of the molecules are negatively charged and the attractive forces between the molecules are to some extent cancelled by repulsive electrostatic forces.

Chlorinated hydrocarbons are good grease solvents. This suggests that the Van der Waals bonds that they can form with *other* molecules, which do *not* contain a lot of halogen atoms, are of much the same strength as those of hydrocarbons.

Halogen-substituted hydrocarbons are to be preferred to hydrocarbons of similar volatility (such as petrol) for a very important practical reason – they are not flammable and so are safer to have around in large tanks. However, most chlorinated hydrocarbons decompose to phosgene, $COCl_2$, a poisonous gas, if they are heated in a flame. Even a cigarette is hot enough.

The high volatility of the chlorinated hydrocarbons used as dry cleaning solvents is useful in that the cleaned textiles can be dried easily by passing warm air through them. (White spirit, for example, is much less volatile and it takes much longer to dry off after treatment.) But a major problem about volatile solvents is that because they evaporate rapidly a high concentration builds up quickly in the air around the table.

The density of a vapour is related to the molecular mass of the compound. The higher the mass the greater the density. If the molecular mass is much greater than 30 the vapour will tend to sink to the floor as it is heavier than air (N_2, mol. mass = 28; O_2, mol. mass = 32). The vapour of 1, 1, 1 trichloroethane (Genklene, mol. mass = 133) is three times as dense as that of ethanol ("meths", mol. mass = 44), although their boiling points (volatilities) are nearly the same. This means that although ethanol vapour may be sucked up into an overhead fume hood, the heavier vapour probably will not. Ideally chlorinated hydrocarbons should be used in a completely enclosed apparatus, like those used in dry-cleaning shops.

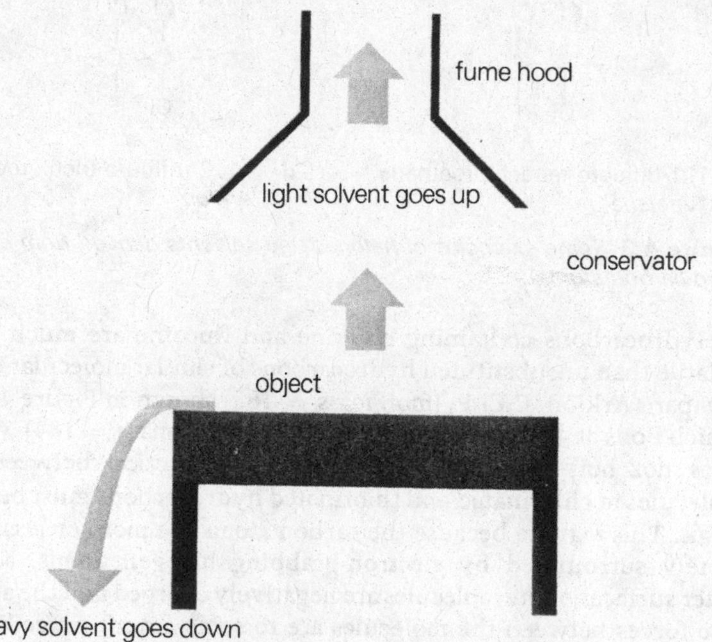

Figure 4.4 *The paths of evaporation for solvents, in use on a conservator's workbench.*

B3 Polar solvents containing oxygen

molecular polarity **Polarity** in molecules is caused by uneven distribution of electric charge. The origin of this imbalance is the great need displayed by elements to achieve an outer shell containing eight electrons when they form covalent bonds (see Book 1). Oxygen is notable for this behaviour and several commonly used polar solvents contain oxygen combined in different ways. Figure 4.5 summarises these:

Name	General Formula	Example
Alcohols	$R-O-H$	C_2H_5OH ethanol
Ethers	$R-O-R$	$C_2H_5-O-C_2H_5$ diethyl ether
Aldehydes	$\begin{matrix}R-&C&=O\\ &\vert&\\ &H&\end{matrix}$	$\begin{matrix}CH_3-&C&=O\\ &\vert&\\ &H&\end{matrix}$ acetaldehyde
Ketones	$\begin{matrix}R-&C&=O\\ &\vert&\\ &R&\end{matrix}$	$\begin{matrix}CH_3-&C&=O\\ &\vert&\\ &C_2H_5&\end{matrix}$ methyl ethyl ketone
Acids	$\begin{matrix}R-&C&=O\\ &\vert&\\ &O-H&\end{matrix}$	$\begin{matrix}CH_3-&C&=O\\ &\vert&\\ &O-H&\end{matrix}$ acetic acid
Esters	$\begin{matrix}R-&C&=O\\ &\vert&\\ &O-R&\end{matrix}$	$\begin{matrix}CH_3-&C&=O\\ &\vert&\\ &O-C_2H_5&\end{matrix}$ ethyl acetate

Figure 4.5 *Classes of oxygen-containing organic substances. (The symbol R– stands for a hydrocarbon group, such as methyl, ethyl, and so on).*

A high proportion of the polar organic solvents used in conservation fall within the categories shown in this table. The range of slight variations of composition allow a very precise selection to be made. The polarity varies between the different classes (as will be explained shortly) and can be adjusted *within* these classes by changing the radicals (denoted R–). Acids, however, will not be discussed further until Chapter 6.

Even more variations are available if you use solvents with *two* oxygen containing groups. These groups may differ from one another as, for example, 2-ethoxy-ethanol, $C_2H_5-O-C_2H_4-OH$, which is *both* an alcohol and an ether.

In all these molecules, the oxygen atom acquires the major share of electrons in the covalent bonds which it forms, and the oxygen atom therefore becomes a *negatively* charged region of the molecule. The distinctions between the classes derive from the *sources of the electrons* which the oxygen atom is acquiring in order to fill its outer shell.

The oxygen atoms in water molecules can only take electrons from hydrogen which is the only other species of atom present.

Hydrogen atoms possess only one electron but if hydrocarbon groups replace hydrogen a more prolific source of electrons becomes available and the resultant *positive* charge, in regions that are deficient in electrons, can then be spread out over several atoms.

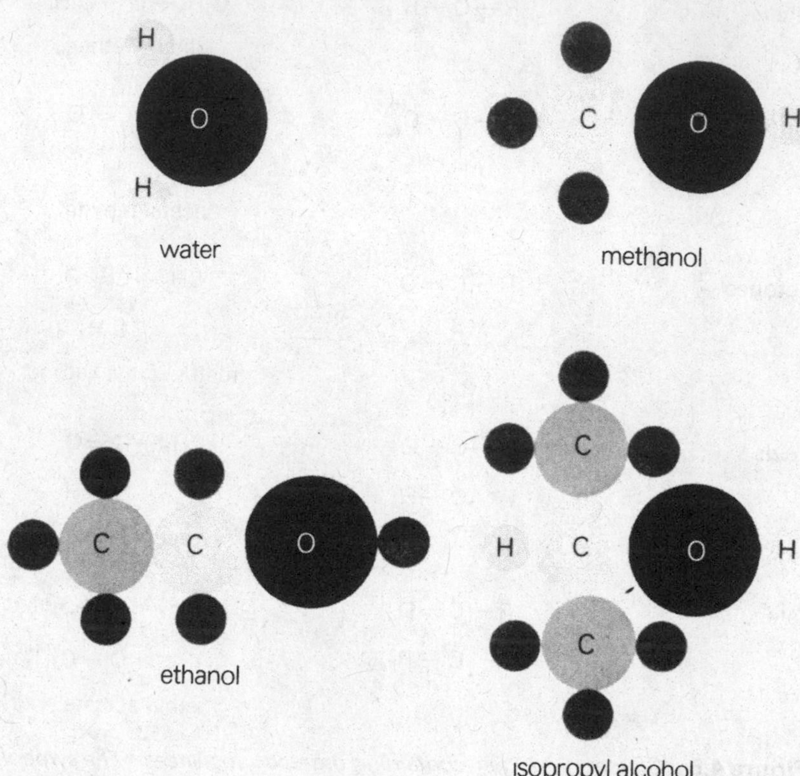

Figure 4.6 *Charge distributions in some polar molecules.*

The consequence of spreading out the positive charge over several atoms is that the polar effects are less pronounced in molecules with longer hydrocarbon chains. Notice in particular that hydrogen atoms *not directly* attached to the oxygen are unaffected. Since these hydrogens are *not* starved of electrons, they are unable to form the hydrogen bond bridges to other molecules.

Alcohols

Hydrogen atoms that *are* joined directly to the oxygen, however, *do* suffer some loss of electrons and in alcohols, therefore, there is some degree of *hydrogen bonding*.

alcohols

Because of these hydrogen bonds, **alcohols** can be expected to mix well with water. In molecules with large hydrocarbon chains the non-polar part of the molecule is not compatible with water and so miscibility (the ability to mix) with water is less for higher alcohols. The *higher* alcohols (those with higher molecular mass) have longer and heavier molecules which are more likely to become entangled with each other.

It follows that these liquids are more viscous than the familiar alcohols like ethanol (IMS, meths) and propan-2-ol (IPA). More energy is needed to separate the molecules and disperse them into the air and so they have higher boiling points, low volatility and are slow to evaporate.

Ketones and aldehydes

carbonyl group

The characteristic group is the **carbonyl group,** C=O. The double bond shows that there is a high density of electrons between the atoms and of course the oxygen gets the greater share of them. The O atom is therefore charged negative and the C, positive. Compounds containing carbonyl groups are very polar. The polarity is higher or lower depending upon what groups are attached to the carbon. If one of the spare carbon bonds is occupied by hydrogen the substance is called an **aldehyde.** Formaldehyde is the most polar of this class:

aldehyde

$$\mathrm{(H)(H)}\overset{+}{\mathrm{C}}=\overset{-}{\mathrm{O}}$$

The attraction between molecules is so great that formaldehyde is rarely found as *monomer* (single molecule) but as *polymers* made from groups of molecules.

Acetaldehyde and acetone are slightly less polar:

$$\mathrm{(CH_3)(H)C=O}$$

acetaldehyde

$$\mathrm{(CH_3)(CH_3)C=O}$$

acetone

As bigger hydrocarbon groups are used, the positive charge becomes diffused over many atoms and the polar effects decrease slightly.

There is no possibility of hydrogen bonding between these molecules because their hydrogen atoms are not joined directly to oxygen and are not starved of electrons. The negatively charged oxygen atom is, however, an ideal site for hydrogen bonding from other types of molecules. Thus acetone mixes with water, and may be regarded as a reasonable solvent to use on a hydrogen-bonded solute, such as nitrocellulose glue.

Because acetone is miscible with water, it can be used to speed up the drying of water from wet objects. It evaporates very fast, however, which cools the surface, and this can make water vapour from the air condense, causing bloom. Bloom effects may also be due to varnish redepositing in a finely divided form from solution in the acetone. If this occurs, a **ketone** with a higher molecular mass, and hence slower evaporation rate, would probably be more successful.

ketone

Another useful solvent is diacetone alcohol (4-hydroxy-4-methyl pentan-2-one):

$$CH_3-\overset{\displaystyle O}{\underset{}{\overset{\|}{C}}}-CH_2-\overset{\displaystyle CH_3}{\underset{\displaystyle OH}{C}}-CH_3$$

It is slow to evaporate and can form hydrogen bonds through its – OH group. The combined alcohol and ketone groups make this compound a good solvent for a wide variety of resins.

Ethers

ethers The general formula for **ethers** is: R – O – R, where R – stands for an hydrocarbon group and the two Rs may be different from each other. The oxygen atom, again, becomes negative but the positive charge is distributed thinly over both R – groups, which means that the polar effects are weak. For this reason, ethers are volatile and are not very miscible with water, though they will dissolve if a small amount of alcohol is present. Ethers are good solvents but evaporate too quickly to be useful in removing varnishes. They are sometimes used for a quick surface clean on a solvent sensitive surface, where their quick evaporation prevents them penetrating, swelling or dissolving the surface finish. They must always be used with care because of their extreme flammability and, as all ethers have anaesthetic properties, fume extraction should be used.

Esters

esters The general formula for **esters** is:

$$R-\underset{\displaystyle O}{\underset{\|}{C}}-O-R$$

Although they do not form hydrogen bonds with their own molecules, they do hydrogen bond with water. They are polar and miscible with water. Examples in conservation use include:

methyl acetate	CH_3COOCH_3	fast-evaporating
ethyl acetate	$CH_3COOCH_2CH_3$	
n butyl acetate	$CH_3COOCH_2CH_2CH_2CH_3$	
amyl acetate	$CH_3COOC_5H_{11}$	
2 ethoxy-ethyl acetate	$CH_3COOCH_2CH_2OCH_2CH_3$	slow-evaporating

These are the best solvents available for nitrocellulose; the esters with higher molecular mass are also good for polyvinyl acetate and polymethylmethacrylate (Perspex).

The three major divisions of organic solvents described in this section have been chosen to exemplify the general principles outlined at the start of the chapter. Quite incidentally they probably cover almost all organic solvents used commonly in conservation. There are a few exceptions – the selectivity of N-methyl-2-pyrrolidone in dissolving starch, but not cellulose, has already been mentioned. Similarly, little is known about why morpholine is a selective solvent for shellac, casein and oxidised linseed oil.

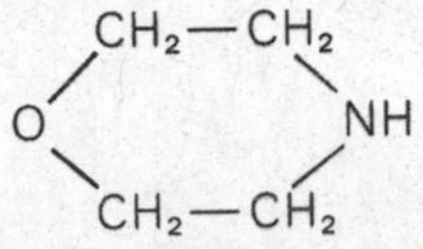

```
   CH2—CH2
  /       \
O          NH
  \       /
   CH2—CH2
```

Figure 4.7 *The structure of morpholine.*

Some of the more powerful solvents (powerful in the sense that they move dirt that other solvents will not touch) may not be forming true solutions. They may be reacting chemically with the dirt. This means that what could be recovered from the solvent would not be chemically identical to what was dissolved.

5

Cleaning with water

A The special properties of water

B Ionic solutions in water

C Purifying water

C1 *Deionisation*

C2 *Water softeners*

D Soaps and detergents

Cleaning with water

Water is the most important liquid cleaning agent, with the triple advantages of being very cheap, easily available and without hazard to the conservator. It is a remarkable substance, different in almost every scientifically measurable respect from other solvents, which is why it has a chapter to itself. This chapter describes some of water's special properties in terms of molecular structure. Its surface tension properties and its ability to dissolve ionic compounds are explained. The conservation implications of its ability to act as a solvent for ionic dirt, and of the need to purify water by removing ions, are also considered.

In practice, water is rarely used alone as a solvent, and all kinds of additives are used to modify its properties. Soaps, detergents, and miscible organic solvents are added for various practical reasons. This chapter explains how these additives work.

A The special properties of water

The special properties of water can be related to its molecular structure, which governs the way its molecules interact with each other and with other substances. You have already learnt that when covalent bonds are formed between two hydrogen atoms and an oxygen atom the electrons are not shared equally. The strong need for electrons shown by oxygen atoms leads to the electrons in the bond being drawn away from the hydrogen atoms onto the oxygen atom. The oxygen atom thus becomes *negatively* charged, leaving the hydrogen atoms *positively* charged. The distribution of electrons around the oxygen atom is such that both the hydrogen atoms are on one side of the oxygen, making the molecule a *dipole* (see Chapter 1). If the molecule were linear, although the actual bonds would be polar, the whole molecule would be non-polar as the effects would be equal and opposite.

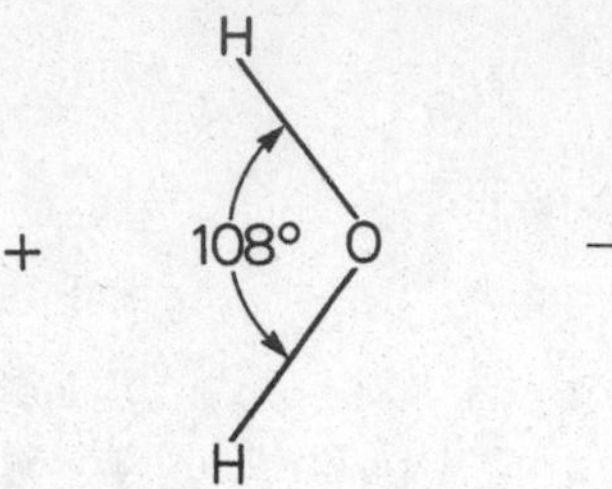

Figure 5.1 *The water molecule.*

The negatively charged oxygen atom provides a site for *hydrogen bonding* between the water molecules. The polarity of the water molecule and its ability to form hydrogen bonds allow water to dissolve, soften or swell organic substances whose molecules contain enough polar groups (often –OH), for example, polyvinyl alcohol, polyethylene glycol and starch, to say nothing of wood, leather and many organic structural materials. The strong polarity of water molecules is one of the reasons why water is also able to dissolve ionic compounds (salts) which most organic solvents do not touch.

The strong secondary bonding between water molecules is manifest in its physical properties too. Extra energy is needed to break the hydrogen bonds and cause the molecules to move fast enough to escape from the surface and disperse into the atmosphere as a gas. This is why water has a boiling point well above that to be expected of a compound with so low a molecular mass (18). Similar compounds which have higher molecular masses – eg hydrogen sulphide, H_2S (34), and carbon dioxide, CO_2 (44) – are gases at ordinary temperatures. Clearly it is useful that water is liquid but other properties of water are not so convenient for cleaning, especially its high surface tension.

The tendency of water molecules to cling to one another, and so to require a lot of energy to move them apart means that water does not flow as easily as other liquids composed of molecules which do not form such strong secondary bonds. Water is viscous compared with ether, acetone and hexane which, although made of heavier molecules, are not hydrogen-bonded, so flow more easily.

B Ionic solutions in water

The most common compounds of metals are their salts, which have ionic bonds. As explained in Book I (Chapter 4), these result from the complete transfer of electrons from metal atoms to non-metallic atoms or groups. The metal atoms become electrically charged *positive ions* and the non-metallic parts become *negative ions*. The conventional symbols for ions are, you will remember, the usual symbols for the atoms or groups of atoms with as many + signs or − signs as electrons lost or gained. Thus Na^+ is a sodium atom which has lost one electron to become a sodium ion. Fe^{2+} is two electrons short of an iron atom, Cl^- has one more electron than a chlorine atom and SO_4^{2-}, the sulphate ion, is an assembly of atoms carrying two more electrons than all the individual atoms would separately. These arrangements satisfy the *rule of eight electrons* to give a complete and stable outer shell of electrons for all the atoms. They also lead to very strong electrostatic attractive forces which make up the primary ionic bond between the positive ions and the negative ions. As these forces act in all directions it is inaccurate to think of "molecules" of salts. Instead, there are regular arrays of vast numbers of ions arranged so that positive ions are surrounded by negative ions. These arrays are recognised as *crystals* (see Book 1, Chapter 4).

Consider how crystals might dissolve in water. It takes a lot of energy to disrupt the crystal structure of an ionic compound and so only the fastest water molecules will succeed in dislodging ions from the crystal. These extra fast moving molecules would soon be used up if there was no way of recovering the energy they spent in knocking ions out of the crystal. Being polar, water molecules can form electrostatic bonds with ions of either sign:

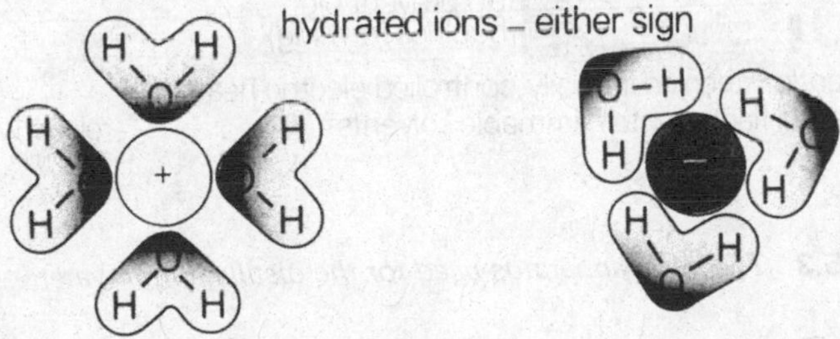

Figure 5.2 *The way water molecules can form electrostatic bonds with both positive and negative ions to become hydrated ions of either sign.*

The ions are then said to be **hydrated** (combined with water). Two consequences follow: **hydrated ions**

- There is a drop in the potential energy of the water molecules. The *potential* energy lost by the water molecules as they cluster around the ions is converted to *kinetic* energy which, after transfer by collisions to unattached water molecules, maintains the supply of molecules moving fast enough to knock more ions out of the crystal.
- The ions coated with water molecules are less likely to recombine with the crystal because they are now the wrong shape, and because the electrostatic forces in the crystal have to act over a longer distance, and are consequently weakened.

C Purifying water

Because water is such an effective solvent for ionic compounds you would expect water taken straight from the tap to contain ions dissolved from the rocks it has percolated through and flowed over. Water from limestone areas will have significant quantities of ions (Ca^{2+}, Mg^{2+}, SO_4^{2-}, Cl^-, CO_3^{2-}, and HCO_3^-) dissolved in it. Such water is said to be **hard**, and you will know that it forms an insoluble scum with soap and does not lather easily. The contrast with **soft** water is well known.

hard and soft water

For conservation use, water needs to be purified. The two methods most often used are **distillation** and **deionisation**. In distillation the water is boiled and the steam that comes off is collected and *condensed* (cooled to form a liquid from a gas) using apparatus which is in principle like that in Figure 5.3.

distillation and deionisation

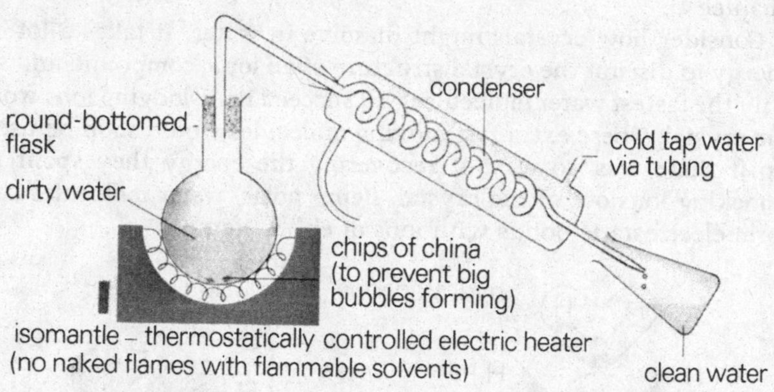

Figure 5.3 *The basic apparatus used for the distillation of water.*

condensate

The **condensate** (that which is condensed) is called distilled water and consists of water molecules only. When a solution is boiled, the light solvent molecules leave the surface of the liquid more readily and evaporate off into the atmosphere, the heavier solute molecules remaining behind. This same process of evaporation also occurs at room temperature, although much more slowly, as there is always a small proportion of molecules with enough energy to leave the liquid. As energy in the form of heat is supplied to the water the hydrogen bonds between the molecules break, and the molecules move about rapidly until they are leaving the surface in considerable numbers. The amount of water vapour in the air above the liquid is measured as the **vapour pressure**. The vapour pressure increases as the temperature of the liquid rises. When the vapour pressure of molecules trying to get *out* of the water equals the pressure of the air keeping them *in*, the water boils. (This relationship between boiling and air pressure is why water boils at a lower temperature on the top of a high mountain, where the air pressure is lower.)

vapour pressure

C1 Deionisation

The deionising columns used in purifying water for conservation contain a mixture of two different resins, one to remove positive ions, the other negative ions. The resins are insoluble and the dissolved ions in water become firmly attached to what are in effect giant cations and anions. Let us call the two resins *Resin – H* and *Resin – OH* and look at the removal of calcium sulphate as an example:

On the first resin:

$$\text{Ca} + \text{SO} + \underset{\text{acid resin}}{\text{Resin–H}} \rightarrow \text{H}^{+} + \text{SO}_4^{2} + \underset{\text{spent resin}}{\text{Resin–Ca}}$$

On the second:

$$\text{H}^{+} + \text{SO}_4^{2} + \underset{\text{base resin}}{\text{Resin–OH}} \rightarrow \underset{\text{pure water}}{\text{H}_2\text{O}} + \underset{\text{spent resin}}{\text{Resin–SO}_4}$$

(You should note that no attempt has been made to balance these equations.) When all the sites at which reaction can take place on the resin are used up the resin has to be regenerated so that it can be used again. The spent cartridges of resin are usually returned to the manufacturer or supplier for regenerating.

Deionisation is fairly economical in soft water areas where there is not much ionic material to take out and one charge of resin lasts a long time. In areas of hard water the resins get used up quickly and frequent replacement is necessary. To determine when the resin cartridges need replacement, the equipment is fitted with a conductivity meter working either from a battery or from the mains. This passes electricity through a sample of the water as it runs through the system and measure the ability of the sample to conduct electricity. Pure water has a low **conductivity** because conduction depends upon ions to carry the electricity. When even a small quantity of salts dissolved as ions is present, the conductivity shoots up and the indicator shows this, usually on a meter dial, indicating that a new resin charge is necessary. It is sometimes thought that when the electricity is switched off the resins will not clean the water. This is not so. The confusion is caused, perhaps, by the explanations involving ions and charges in the water. The electricity supply is only used to *test* the water quality and so with the electricity switched off you cannot tell whether the resins are working or not.

conductivity

C2 Water softeners

Water which has been through a water softener (such as the Permutit type) is *not* deionised; it still contains dissolved salts. This could be shown by measuring its conductivity. The water has been made "soft" by exchanging the calcium ions, Ca^{2+}, and magnesium ions,

Mg^{2+}, for sodium ions, Na^+. "Softened" water will not produce scum when used with soap as the scum is insoluble calcium and magnesium salts. (The equivalent sodium salts are soluble.) While this treatment is satisfactory for domestic purposes, for conservation uses it is *not*, because salts will still be deposited when this water dries out.

The ion exchange materials in water-softeners are either natural clay minerals (such as sodium aluminium silicate) or beads of synthetic polymer, carrying resins of the structure *Resin – Na* on their surfaces. The reaction (which is not shown as a balanced equation) is:

$$\text{Resin—Na} + Ca^{2+} \rightarrow \text{Resin—Ca} + Na^+$$

solid left in apparatus — different ions in water

When all the active reagent has been used up the reaction can be reversed, by passing a very concentrated solution of sodium chloride through the apparatus. This regenerates the activity of the water-softener.

Treatment of water by the batch using "Calgon" (sodium hexametaphosphate) also softens without deionising. The calcium ions are removed from solution and replaced by sodium ions.

D Soaps and detergents

The advantages of water as a solvent do not include an ability to dissolve grease. Soaps and detergents are additives which enable water to be used as a grease solvent. As you have seen, organic solvents offer an alternative treatment for greasy dirt. Consequently you have to decide when to use "dry cleaning" methods using organic solvents and when to "wash" using water plus an additive. To be able to make this sort of decision, you need to know something about the chemical composition of soaps and detergents and how they work.

soaps Chemically, **soaps** are salts of organic acids, part of whose molecule is a long hydrocarbon chain. These acids in other combinations are constituents of animal fats and vegetable oils and are collectively known as **fatty acids**. Oleic acid, $C_{17}H_{33}COOH$, will serve as an example. If you compare this molecule with acetic acid, CH_3COOH, you will see that the difference is that the acid-making part of the molecule

fatty acids

$$—\underset{\underset{O}{\|}}{C}—O—H$$

(called carboxyl) is joined to a *long* hydrocarbon group, $C_{17}H_{33}-$, in oleic acid but only a *short* one, CH_3- in acetic acid. When fats are boiled up with **alkali**, salts of the acids are formed. Typical is potassium oleate, a common soap

$$C_{17}H_{33}-\underset{\underset{O}{\|}}{C}-O-K$$

which dissolves in water forming positive K^+ ions and negative oleate ions $C_{17}H_{33}COO^-$. It is these negative ions which make the soap work.

The oleate hydrocarbon chain is *unsaturated*; it has a double bond in the middle. Ordinary soap is a mixture of the sodium salts of the saturated fatty acids which have an odd number of carbon atoms in the hydrocarbon chain, from octoic, $C_7H_{15}COOH$, to stearic, $C_{17}H_{35}COOH$. The sodium salts are harder and less soluble than the corresponding potassium salts. Potassium salts derived from oils such as potassium oleate are called soft soaps.

In the long hydrocarbon chains the orientation of the bonds around each carbon atom is tetrahedral (see Book I, Chapter 4).

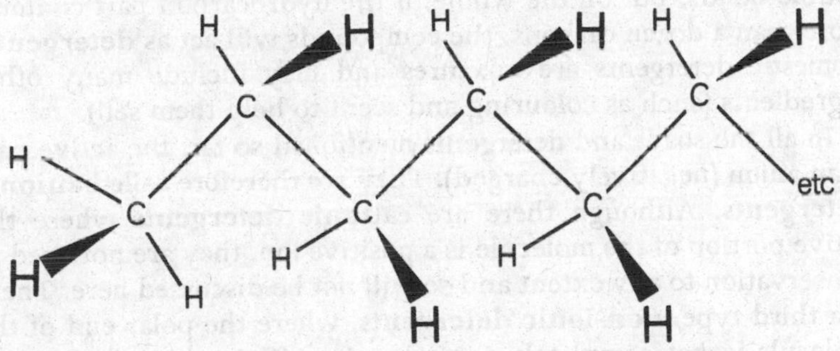

Figure 5.4 *A structural representation of a hydrocarbon chain.*

This zig-zag structure of the carbon chain is often merely represented as a wriggly line, for example:

$$\sim\sim\sim\underset{\underset{O}{\|}}{C}-O^-K^+$$

The negative ion consists of a water loving, *hydrophilic* part at the charged end with high electron density due to the oxygen atoms and a *hydrophobic*, water hating, long chain hydrocarbon at the other. This acts like other hydrocarbons such as oils, fats and waxes which are also hydrophobic. A convention sometimes used to indicate this is to draw the molecule like a tadpole showing the hydrophobic tail and hydrophilic head:

$$\sim\sim\sim\bullet$$

One end of the ion would prefer to dissolve in water, the other end is more compatible with grease.

Detergents generally have molecules with the same type of structure but are derived from a mineral acid, such as sulphuric acid, H_2SO_4, or phosphoric acid, H_3PO_4. The *sulphonate* detergents have the structure:

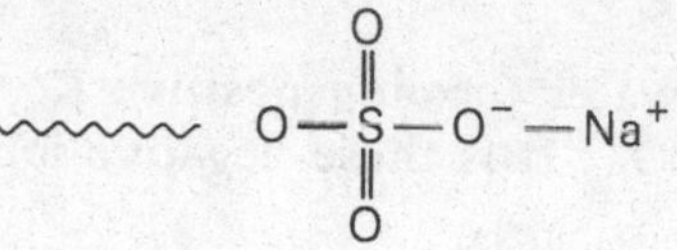

$$\sim\sim\sim\;O{-}\overset{\displaystyle O}{\underset{\displaystyle O}{\overset{\|}{\underset{\|}{S}}}}{-}O^{-}{-}Na^{+}$$

There is a variant which has SO_3 rather than SO_4 in the chain and includes a benzene ring:

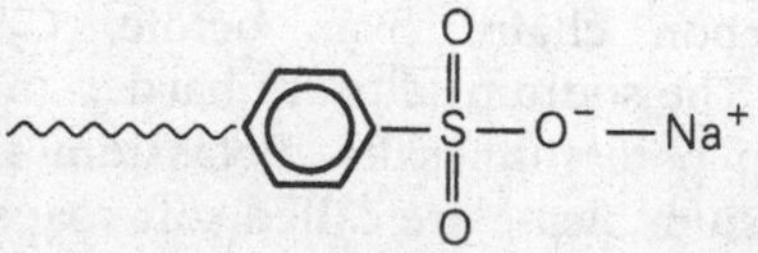

$$\sim\sim\sim C_6H_4{-}\overset{\displaystyle O}{\underset{\displaystyle O}{\overset{\|}{\underset{\|}{S}}}}{-}O^{-}{-}Na^{+}$$

Chemical variations, as you can imagine, are endless, with shorter or longer chains, branched or straight ones, with or without rings or double bonds; but on the whole, if the hydrocarbon part contains more than a dozen carbons, the compounds will act as **detergents.** Domestic detergents are mixtures and may include many other ingredients (such as colouring and scent to help them sell).

detergents

In all the soaps and detergents mentioned so far, the active part is an anion (negatively charged). They are therefore called **anionic detergents**. Although there are **cationic detergents** where the active portion of the molecule is a positive ion, they are not used in conservation to any extent and so will not be discussed here. There is a third type, **non-ionic detergents**, where the polar end of the molecule is not completely ionic but is sufficiently polar to give solubility in water. These are extensively used. One of these occurs naturally in the plant soapwort and has been successfully used in conservation for washing textiles. Just as safe, cheaper and more consistent in properties are the industrially made ones, known (from their mode of manufacture) as ethylene oxide condensates. The general pattern of their molecules is shown below.

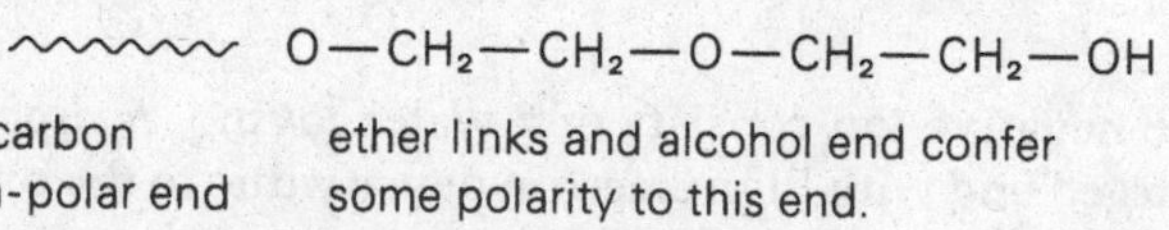

anionic and cationic detergents

non-ionic detergents

$$\sim\sim\sim\; O{-}CH_2{-}CH_2{-}O{-}CH_2{-}CH_2{-}OH$$

long hydrocarbon chain – non-polar end; ether links and alcohol end confer some polarity to this end.

Variations in the hydrocarbon chain, in the spacing of the oxygen atoms and in what spaces them out (sometimes benzene rings) form the many different members of this class of detergents.

Like substances dissolve like. The polar end of the molecule gives it solubility in water and the non-polar end gives solubility in grease. Grease molecules become attached to the tails (the hydrocarbon chains) by Van der Waals forces and are carried into solution. Let us look at this process in more detail.

The action of soaps and detergents at the interface between water and a greasy surface explains why these additives are known as **surfactants**. Certainly the surface of the water should be expected to be covered with the active ions, the hydrophobic tails sticking out from the water and able to get at the grease (Figure 5.5).

surfactants

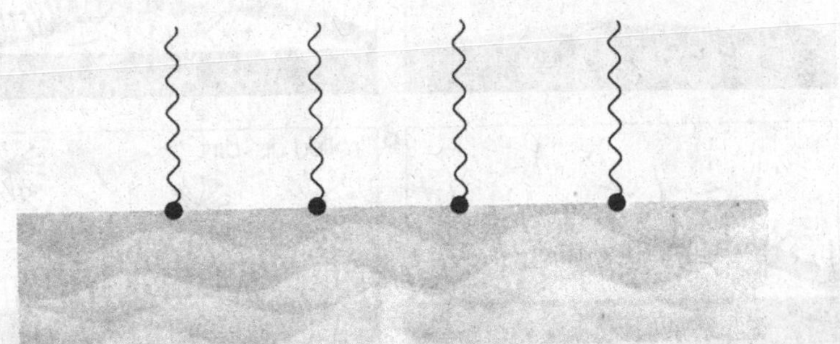

Figure 5.5 *The hydrophobic tails of a surfactant sticking out from the water surface; the hydrophilic heads remain in close contact.*

Within the body of the liquid the polar anions form **miscelles**, small clusters with the hydrophobic tails all together *inside* a rough sphere and the hydrophilic ends on the outer surface in contact with the water. "Like with like" yet again.

miscelles

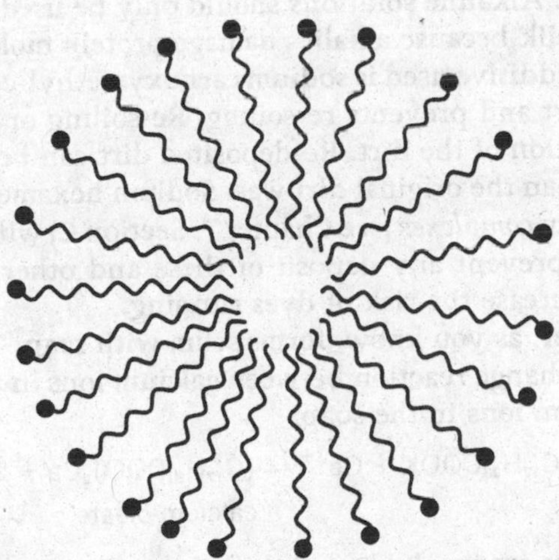

Figure 5.6 *A cross-section through a miscelle which is roughly spherical in shape.*

The centre of this miscelle is a non-polar solvent. Grease molecules can be absorbed and held in solution in these centres, thus removing them from the object or surface. In addition the dirt may be removed in larger lumps consisting of many grease molecules.

The following diagram shows how a washing solution containing a surfactant, gets into, lifts away and suspends the dirt:

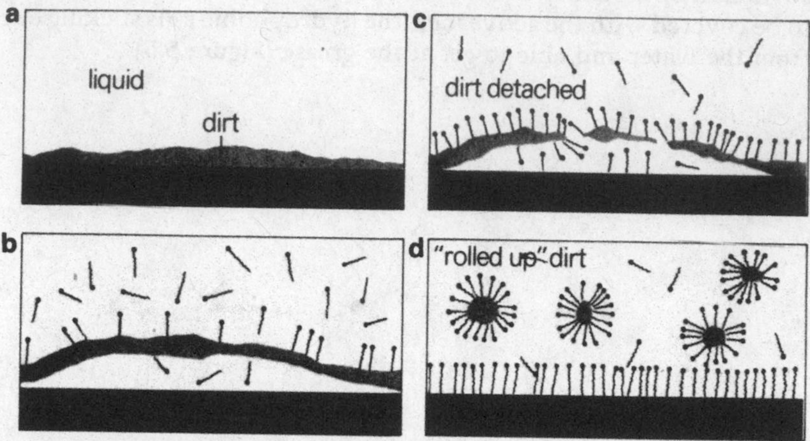

Figure 5.7 *A series of illustrations to show the process by which a surfactant contained in a washing solution acts on dirt.*

In addition to surfactants, water washing solutions may contain compounds to keep the solution mildly alkaline. Sodium carbonate is used with sodium silicate (also called sodium metasilicate) to stabilise the pH (see Chapter 6). Alkaline solutions clean more effectively, and help the surfactants to attack grease. They prevent the formation of insoluble metal salts or soaps which would be precipitated. Alkaline solutions should only be used on cotton, not on wool or silk because alkalies damage protein molecules.

Another additive used is sodium carboxymethyl-cellulose, which suspends dirt and prevents re-soiling. Re-soiling or greying is due to re-deposition of the dirt. Re-deposited dirt can be more difficult to remove than the original dirt was. Sodium hexametaphosphate is added to form *complexes* (see Chapter 7, Section C) with calcium salts in order to prevent any deposit of these and other metals. It can however, increase the risk of dyes running.

Hard water, as you know, forms scum with soap. This is because of an ion exchange reaction between calcium ions in the hard water and potassium ions in the soap:

$$2C_{17}H_{33}COOK + Ca^{2+} \rightarrow \underset{\text{calcium oleate}}{(C_{17}H_{33}COO)_2Ca} + 2K^+$$

Calcium soaps are insoluble and these form the scum. When soap is chosen for washing delicate textiles, deionised water should always be used. Even where deionised water is used to make up the washing solution there may be calcium salts in the object which would react, so that an insoluble scum is left behind on the object. For this reason, soaps of this type are hardly ever used in conservation work. The use of soft soap on marble or other calcareous stones should also be avoided because it forms a yellowish-grey, insoluble film that is difficult to remove. The immediate advantage of the synthetic detergents is that their calcium salts are soluble and do not form scum.

It is important to wash away all the detergent and other additives with clean fresh water so that no residues are left behind. It is difficult to predict exactly how any residues might react with the material of which the objects are made over a long period of time, and it has been observed in the laundry industry that detergent residues increase the rate of resoiling. The object will get dirty again more quickly if a residue is left.

The strong polarity of detergents has two disadvantages which cannot always be avoided: swelling of the material being washed (eg wood, ivory, paper, leather, textile fibres, etc.) which may be harmful in itself or may encourage dyes to bleed, pigment to lift for example; and the difficulty of removing the detergent from the object after cleaning.

6

Water, acidity and alkalinity

Water, acidity and alkalinity

There are many conditions of objects, and treatments for them, where acidity and alkalinity are important factors. This chapter explains what makes some solutions in water acidic and others alkaline. The chapter also discusses what the pH scale of measurement means as well as looking at ways of measuring and controlling pH.

A Hydrogen ions in water

A1 Molecular conditions in pure water

A further consequence of hydrogen bonding between water molecules has to be considered in order to understand acids and alkalies. Figure 6.1 shows two molecules of pure water joined together by a hydrogen bond:

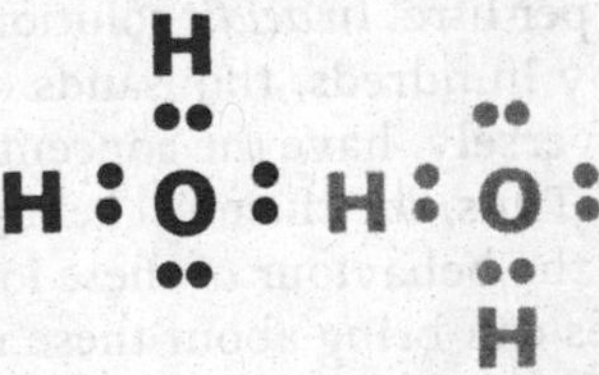

Figure 6.1

The hydrogen atom forming the bridge is involved in two bonds, a covalent bond to its own oxygen atom and a hydrogen bond to the oxygen of the nearby molecule. However, the electron arrangements are exactly the same in both bonds. The only distinction between them is their length. As the molecules jostle around, first one bond is the longer, then the other. In collisions which are hard enough to

break a bond, the longer, weaker one is the more likely to break. Sometimes the pair will break to form two separate H_2O molecules; at other times it will break to H_3O and OH. In the second case the H_3O has gained the extra positive charge of the hydrogen nucleus and so H_3O is a positive ion, H_3O^+. The molecule which has lost a hydrogen nucleus still holds its electron and so the ion OH^- is formed. In summary, what has been described is the reaction:

$$2H_2O \rightarrow H_3O^+ + OH^-$$

It is, of course, quite possible that each of these ions will bump into others of the opposite sign as they wander around in the liquid. Then the reverse reaction may occur:

$$H_3O^+ + OH^- \rightarrow 2H_2O$$

The chances of this reverse reaction occurring depend upon the population of ions of each sort. A steady state will be reached where the chances of new ions being formed are equal to the chances of the existing ones recombining. The balanced situation where ions are made and destroyed at the same rate is known as a condition of

chemical equilibrium

chemical equilibrium. The chances of ions forming depend upon *temperature* since this controls the probability of bond-breaking collisions. The chances of them recombining depend upon *concentration*, since this controls the probability of positive and negative ions finding each other. The concentration of the two sorts of ion must be exactly the same because an H_3O^+ ion can only be made simultaneously with an OH^- ion.

In pure water at 25°C equilibrium exists with a concentration of one ten-millionth of a mole per litre of each sort of ion. Although this is a *tiny* concentration it does mean that even the purest of pure water is not, chemically, a single molecular species. Moreover, because the ions are chemically more reactive than their parent molecules, their presence strongly influences the chemical interaction of water with other substances.

A chemical equilibrium, like other forms of equilibrium or stability, can be upset by suitable external influences. The conditions of

acidity and alkalinity

acidity and **alkalinity** are just this. The equilibrium is disturbed so that the concentrations of H_3O^+ ions or OH^- ions are no longer one ten-millionth of a mole per litre. In *acidic* solutions the concentration of H_3O^+ is increased by hundreds, thousands or millions of times. *Alkaline* solutions, conversely, have the concentrations of OH^- ions dramatically increased. Thus, the chemical behaviour of the solution becomes controlled by the behaviour of these ions. The compounds called acids and alkalies can bring about these remarkable changes in water when they go into solution.

A2 The pH scale for hydrogen ion concentrations

The concentration of H_3O^+ and OH^- ions in pure water is one ten-millionth of a mole per litre. Written as a fraction this is $\frac{1}{10,000,000}$ which can be written more compactly as 10^{-7}, to be read as "ten to the minus seven".

The convention for describing numbers like this is simply to count how many noughts there are in the number. Numbers bigger than 1 are given a plus index; thus 1000 is 10^{+3}, ie "ten to the plus three" (normally just 10^3 or "ten to the power of three"). Fractions are indicated with a minus index. The fraction $\frac{1}{1000}$ is 10^{-3}, "ten to the minus three".

It is long-winded to refer to concentrations in moles per litre when the numbers become awkward mouthfuls like "one ten-millionth" so a shorthand convention based on the "ten-to-the-something" system has been adopted. When used for describing acids and alkalies this is known as the **pH scale** and describes the *concentration of hydrogen ions* (more strictly, of H_3O^+ ions). **pH scale**

The pH of a solution of concentration 10^{-7} moles/litre is 7, so you can see the origin of the pH number. It is the power of ten which describes the concentration, but with the sign changed. You will have come across the pH scale before and probably know that acid solutions are described by pH numbers *less* than 7. Thus pH6 is acidic. Working backwards you will see that this means a concentration of 10^{-6} mole/litre for H_3O^+ ions. 10^{-6} means $\frac{1}{1,000,000}$ or one millionth.

Now notice this carefully! A concentration of one millionth of a mole per litre is *ten times more* than that of a ten-millionth mole/litre, so a pH6 solution is ten times more concentrated in H_3O^+ ions than is a pH7 solution. Similarly you can show that pH5 is ten times more concentrated than pH6, 100 times more than pH7, and so on. Acids are thus solutions in which the concentration of H_3O^+ is much higher than 10^{-7} mole/litre.

The most immediate consequence of getting more H_3O^+ ions into solution (you will see how this is done, soon) is on the population of OH^- ions. Formation of these ions, by the break-up of H_2O molecules in collisions, still occurs at the same rate as in pure water but now the chances of meeting an H_3O^+ ion and recombining with it are vastly increased so the population of OH^- ions collapses. So when water is made acidic not only are there more hydrogen ions (hydrated to H_3O^+) but also many fewer hydroxide ions. So the ratio of H_3O^+ to OH^-, which is 1:1 at pH7, changes dramatically. Figure 6.2 shows the ratio of populations for various pH values. **ion ratios**

pH	Ratio of H_3O^+ ions to OH^- ions	Moles/litre of H_3O^+
7	1	one ten-millionth
6.85	2	
6.65	5	
6.5	10	
6	100	one millionth
5.5	1000	
4	1 million	one ten-thousandth
2.5	1 thousand million	
1	1 million million	one tenth

Figure 6.2 *pH from 1 to 7*

Figure 6.2 concentrates upon weakly acid solutions. Very small changes in pH mean larger changes in the ratio of H_3O^+ to OH^-. pH6 may not sound very different from pH7, but you can see that already the H_3O^+ ions are the dominant species in the solution. The pH description of alkaline solutions uses numbers greater than 7. This will be dealt with later, but let us see what sort of substances cause these remarkable changes in water when they go into solution.

B Acids, bases and alkalies

B1 How water becomes acidic

You have seen how ionic crystals made up of separate ions go into solution in water *because* the polar water molecules offer electrostatically "comfortable" places for the ions to go. The pre-formed ions, such as Na^+ and Cl^-, exist by the exchange of an electron between a metal atom which is happy to lose it and a non-metal anxious to grab it. Similar compounds with hydrogen in place of the metal are covalent but strongly polar, with the hydrogen *almost* ionised, eg $\overset{+}{H} - \overset{-}{Cl}$. Put this in water and the hydrogen finds a better share of electrons if it goes on to a water molecule. Thus:

$$HCl + H_2O \rightarrow H_3O^+ + Cl^-$$

hydrogen chloride and water → hydrochloric acid

or, showing the electrons:

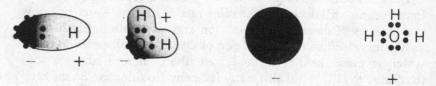

Figure 6.3

Now, different kinds of molecules may ionise to a greater or lesser extent as they dissolve in water. At the extremes, common salt, an ionic compound, dissolves entirely to become separate ions; sugar, a covalent compound with –OH groups available for hydrogen bonding, does not ionise at all in solution. These two statements could be checked by electrical conductivity tests on solutions. The amount of ionisation in solution is closely linked with the polarity in the undissolved molecules. Acids, like water, have strongly polar molecules with hydrogen atoms as the positive part and oxygen atoms are very often the source of the polarity as they draw electrons to themselves. Apart from hydrochloric acid and hydrofluoric acid, HF, all the common acids contain oxygen and are known collectively as oxyacids. Here are some of their molecular structures.

oxyacids

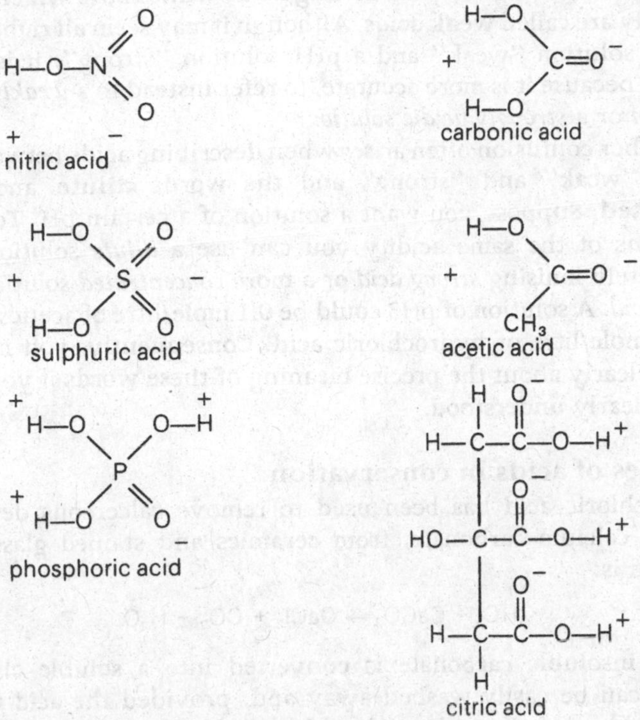

Figure 6.4 *Some oxyacid structures (± shaded).*

The significant structural common features of these molecules are the OH groups covalently coupled through another atom to a double bonded oxygen atom. The electrons are pulled towards the double-bonded oxygen atom(s) leaving the hydrogen precariously bonded. The molecules are polarised as shown.

There is strong hydrogen bonding between the molecules of these acids in pure form.* You may have met acetic, sulphuric and phosphoric acids in pure or nearly pure form, and know them to be sticky viscous liquids. The argument for their high viscosity is the same as for water.

Mixing these substances with water gives the loose hydrogen atoms the option of being attached to water molecules as H_3O^+ ions instead of to their original molecule. The option is taken up, sometimes completely, sometimes less than completely, depending upon how loosely the hydrogen atoms are held to the acid molecule. Thus hydrogen chloride and nitric acid ionise completely in water whereas among organic acids only a small proportion of the molecules break up. For acetic acid the figure is about 1 in 1,000, but this is millions of times more than the number of broken water molecules that exist in pure water, so acetic acid successfully floods the solution with H_3O^+ ions.

The fact that some acids ionise in solution more than others leads to some confusion in conversation. Speaking scientifically, acids

* *Carbonic acid cannot be prepared pure because it decomposes to H_2O and CO_2.*

strong and weak acids

which ionise completely are **strong** acids while those which ionise partially are called **weak** acids. Although it may seem all right to call a pH6 solution "weak" and a pH1 solution "strong", it is much better, because it is more accurate, to refer instead to a *weakly acidic solution* or a *strongly acidic solution*.

dilute and concentrated acids

Further confusion often arises when describing acids between the terms "weak" and "strong" and the words **dilute** and **concentrated**. Suppose you want a solution of a certain pH. To make solutions of the same acidity you can use a *dilute* solution of a completely ionising *strong acid* or a more *concentrated* solution of a *weak acid*. A solution of pH3 could be 0.1 mole/litre of acetic acid or 0.001 mole/litre of hydrochloric acid. Consequently, you need to think clearly about the precise meaning of these words if you wish to be clearly understood.

B2 Uses of acids in conservation

Hydrochloric acid has been used to remove calcereous deposits, mainly calcium carbonate, from ceramics and stained glass. The reaction is:

$$2HCl + CaCO_3 \rightarrow CaCl_2 + CO_2 + H_2O.$$

The insoluble carbonate is converted into a soluble chloride which can be easily washed away and, provided the acid is also thoroughly rinsed out there should be little, if any, damage. It has also been recommended for removing lead carbonate from lead objects. This would have to be done very carefully indeed because metallic lead is attacked by hydrochloric acid.

Hydrofluoric acid has been used to remove iron stains. Unlike hydrochloric acid, it does not react with carbonate to liberate CO_2, so has been used on marble and limestone. Also since it is not an oxidising acid, as is for instance sulphuric acid, it does not readily attack the cellulose of cotton textiles or paper so can be used on these (provided it does not affect any colouring matter which *should* be on the object). Nitric acid, however, is a powerful oxidising agent.

Orthophosphoric acid H_3PO_4 is used both to remove iron oxides and to protect iron from corrosion. It probably does both these things by forming *complex compounds*, soluble in the first case, but insoluble in the second. Some comments about complex metal compounds will appear in Chapter 7.

organic acids

Organic acids also find uses. Formic acid, HCOOH, has been used to remove silver chloride from silver and also to dissolve copper corrosion products. Acetic acid CH_3COOH, is sometimes used to prevent dyes running and is often chosen to neutralise excess alkali, after bleaching process for example, but although it may not itself be harmful, residues of the products of reaction in some circumstances can be damaging. Formation of magnesium acetate for example, will tenderise cellulose, promoting oxidative degradation of cellulose as it ages. So the acid and reaction products should be thoroughly rinsed out of the object after use. It attacks lead, and should not be used on this metal.

Citric acid is sometimes used to dissolve cupric oxide, which it does by forming a soluble *complex* (see later). Other organic acids also form complexes and are useful for metal cleaning. Examples are tartaric acid and thioglycollic acid.

tartaric acid

thioglycollic acid

Figure 6.5

B3 pH greater than 7

Now let us turn attention to the other end of the pH scale, 7 to 13. Remember that neutral water, pH7, has the same concentration of H_3O^+ and OH^- ions, and that these concentrations are

$$\frac{1}{10{,}000{,}000} \text{ mole/litre} = 10^{-7} \text{ mole/litre.}$$

Also, we said that pH measures the concentration of H_3O^+, so a pH9 solution, say, has a concentration of H_3O^+ ions of

$$\frac{1}{1{,}000{,}000{,}000} \text{ mole/litre} = 10^{-9} \text{ mole/litre.}$$

This means there is only one hundredth the concentration of H_3O^+ ions as there is in neutral water, so the question is – where have all the H_3O^+ ions gone? The answer is that they have been mopped up by recombination with OH^- ions. To arrange for this to happen the solution has to be swamped with OH^- ions to increase the chances of H_3O^+ ions meeting them and recombining. An obvious source of OH^- ions is a metal hydroxide, caustic soda, NaOH, or slaked lime $Ca(OH)_2$. These are crystalline solids which contain pre-formed OH^- ions. Dissolving such ionic compounds in water therefore raises the OH^- concentration and forces the H_3O^+ concentration down. The ratio of OH^- ions to H_3O^+ ions, which was 1:1 at pH7, shifts rapidly. Compare the following table with Figure 6.2.

hydroxides

pH	Ratio of OH^- ions to H_3O^+ ions	Moles/litre of OH^-
7	1	10^{-7}
7.15	2	
7.35	5	
7.5	10	
8	100	10^{-6}
8.5	1000	
10	1 million	10^{-4}
11.5	1 thousand million	
13	1 million million	10^{-1}

Figure 6.6 *Population ratios for OH^- ions to H_3O^+ ions in solutions with pH greater than 7.*

base

strong bases or alkalies

Any substance which *produces* OH^- ions in solution, whether by virtue of containing them pre-formed or by reactions with water, is known as a **base**. Substances which contain pre-formed OH^- ions in the anhydrous (water-free) condition and which merely release them into solution are **strong bases** or **alkalies**. The word "strong" is used here in the same sense as for "strong" acids – strong bases are completely ionised when dissolved.

The method of producing an alkaline solution is typified by the dissolving of caustic soda (sodium hydroxide):

$$NaOH \text{ in water} \rightarrow Na^+ + OH^-$$

followed by:

OH^-	+	$H_3O^+ \rightarrow 2H_2O$	+	OH^-
large numbers		sparse population		surplus

The H_3O^+ ions are mopped up leaving a surplus of OH^- ions.

Both alkalies and acids can legitimately be regarded as the result of reactions between oxides and water; metal oxides produce alkalies, non-metal oxides generate acids. Example are the slaking of quicklime to make an alkali and dissolving sulphur trioxide to form sulphuric acid:

$$CaO + H_2O \rightarrow Ca(OH)_2 \text{ alkali}$$

$$SO_3 + H_2O \rightarrow H_2SO_4 \quad \text{acid}$$

The difference in the behaviour of these substances when dissolved in more water is an immediate consequence of the metal oxide being *ionic* and the non-metal oxide being *covalent*. Metal atoms give up their outer electrons completely, satisfying the oxygen atoms. So the oxygens make less demands for electrons from the hydrogen atoms. In solution, therefore, $Ca(OH)_2$ breaks into Ca^{2+} ions and OH^- ions. In contrast, sulphur does not shed its electrons to oxygen, so the tendency of oxygen to collect electrons is satisfied at the expense of hydrogen atoms. The molecule H_2SO_4:

O O—H
S
O O—H

therefore, has its weakest covalent bonds at the O—H positions. The ions formed in solution are SO_4^{2-} and H^+, the latter hydrating to H_3O^+.

B4 Weaker bases

Strong bases, (alkalies), find occasional use in conservation, but it is also important to understand how weak bases are effective. We shall now concentrate upon ammonia and its derivatives, and the carbonates of alkali metals (sodium and potassium).

Ammonia and its derivatives

ammonia

When **ammonia** gas, NH_3, dissolves in water three possible molecular states are found which interchange with one another:

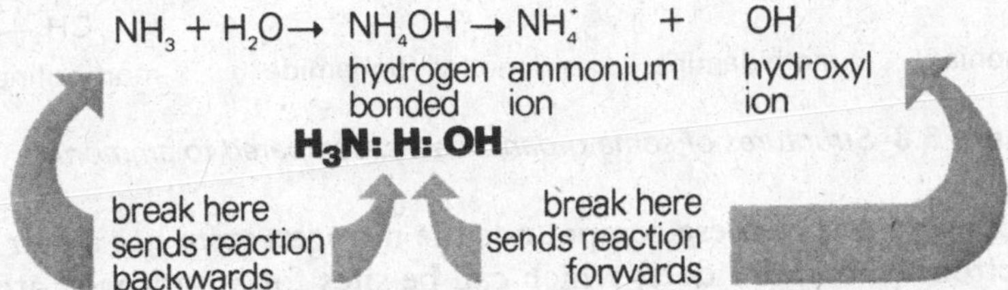

The key to understanding what goes on is the hydrogen bonded compound, NH_4OH (ammonium hydroxide). This forms because both ammonia and water are polar molecules. The case is analagous to the two water molecules shown at the beginning of the chapter which were joined by a hydrogen bond bridge. They could break to form either two separate water molecules or H_3O^+ and OH^- ions.

In ammonium hydroxide the nitrogen and oxygen atoms are joined by a hydrogen bond bridge.

Figure 6.7 *The hydrogen bond bridge between the nitrogen and oxygen atoms in an ammonium hydroxide molecule.*

Either of the arrowed bonds can be broken by collisions and both kinds of break occur. Not only do the two reactions involving the NH_3 grouping have to find equilibrium, but the reaction in which water ionises is involved too, because H_3O^+ ions may be lost as a result of the increase in OH^- concentration caused by the ionisation of ammonium hydroxide. In other words there are three reversible reactions finding mutually stable balance points:

$$NH_3 + H_2O \rightleftharpoons NH_4OH$$
$$NH_4OH \rightleftharpoons NH_4^+ + OH^-$$
$$2H_2O \rightleftharpoons H_3O^+ + OH^-$$

The sign $\rightleftharpoons$ indicates a reaction that is reversible.

The result is, of course, an alkaline solution, rich in OH^- ions and poor in H_3O^+ ions. The strength of ammonia as a base is such that a 0.1 mole/litre solution of ammonia is about pH11. This indicates that about 1% of the ammonia molecules dissolved in the solution are ionised at equilibrium. Notice that the pH value tells us nothing about the state of the other 99% of ammonia molecules – how many are present as NH_4OH for instance.

organic bases

Organic derivatives of ammonia also behave as bases. Some examples are given in Figure 6.8. You can see that their structures are similar to that of ammonia.

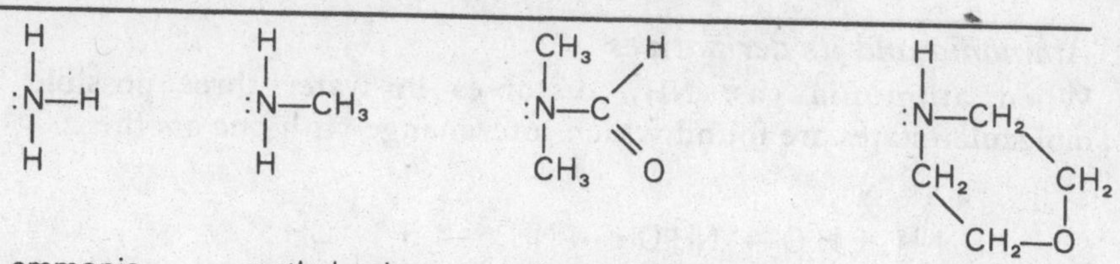

ammonia methylamine dimethyl formamide morpholine

Figure 6.8 *Structures of some organic bases, compared to ammonia.*

In every one of these compounds the nitrogen atom has a pair of electrons (shown by dots) which can be sites for a hydrogen atom which has lost its electron. In water, these compounds will raise OH^- ion concentrations in the same way that ammonia does. The ammonia molecule is not flat. The electronegative nitrogen atom carrying its extra pair of electrons is not in the same plane as the three hydrogen atoms (see Figure 6.9). This is why the molecule is a dipole; ammonia is polar.

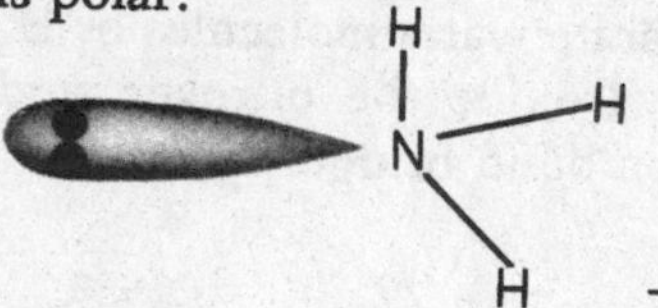

Figure 6.9 *The charge distribution in an ammonia molecule.*

The other nitrogen-containing bases have similar bent structures.

Carbonates

carbonates A chain of reversible reactions similar to that just described for ammonia occurs when carbon dioxide dissolves in water. CO_2, being the oxide of a non-metallic element, produces an acidic solution. Rainwater normally has a pH of about 5 because of the dissolved CO_2. This accounts for much long term deterioration of limestone and marble. (The effects are more drastic in an industrially polluted atmosphere because SO_2 dissolves to reach a much lower pH.) The reactions of CO_2 are:

$$\underset{\text{unreacted solution}}{CO_2 + H_2O} \rightleftharpoons \underset{\text{carbonic acid (unstable)}}{H_2CO_3} \xrightleftharpoons{+H_2O} \underset{\text{bicarbonate ion}}{HCO_3^-} + H_3O^+ \xrightleftharpoons{+H_2O} \underset{\text{carbonate ion}}{CO_3^{2-}} + 2H_3O^+$$

These three reactions and the ionisation of water reaction have to come to equilibrium simultaneously. Measurements show that the concentrations of CO_2 and HCO_3^- are much greater than of H_2CO_3 and CO_3^{2-}.

It is interesting to consider what happens if a soluble carbonate, for example, washing soda, $Na_2CO_3.10H_2O$, is dissolved in water. The equilibrium concentration for the carbonate ion is *low* while for the bicarbonate ion it is *high*, causing:

$$CO_3^{2-} + H_2O \rightarrow HCO_3^- + OH^-$$

to occur which means that sodium carbonate is a *base*. Sodium *bi*carbonate is also a base because if the solution is overloaded with HCO_3^- ions:

$$HCO_3^- \rightarrow CO_2(g) + OH^-$$

occurs, with the CO_2 becoming a gas (shown by the "g") and fizzing out of solution.

Soluble carbonates and bicarbonates, therefore, offer you a source of weak bases which are safer to handle and more gentle to use for conservation purposes than the caustic hydroxides.

B5 Uses of alkalies in conservation

neutralisation

One use of alkalies in conservation is to **neutralise** acids. The action is encapsulated in the adage you may have heard at school – "acid plus base gives salt plus water". This tells you the overall effect of acid meeting base and warns that when you have neutralised an acid there is a salt of some kind to be removed from an object treated in this way. It is worth realising what is happening in the solution. If, for example, sulphuric acid is being neutralised by a solution of sodium hydroxide, the solution contains H_3O^+ and SO_4^{2-} ions from the acid and Na^+ and OH^- ions from the alkali. The only reaction occurring is:

$$H_3O^+ + OH^- \rightarrow 2H_2O$$

caused by the high population of both ions, which increases their chances of meeting and recombining. As the solution dries out, H_2O molecules evaporate raising the concentration of all the ions and causing more recombination of H_3O^+ and OH^- ions. Eventually, if *exactly* the right amount of alkali is present, only Na^+ and SO_4^{2-} will remain and will combine to form solid crystals. It is imperative, even if the neutralisation is *exactly* done, to wash away the saline residues before the object is dried. In fact you would not choose strong bases. The alkaline hydroxides are rarely used for **deacidification** because they might cause further damage to the object by overcompensating for acidity and because unduly rapid reactions could cause damage by releasing a lot of heat.

deacidification

The favoured base for many deacidifying tasks is ammonia solution, partly because ammonia is a weak base, but also because most ammonium salts are very soluble.

Ammonia has been used to deal with the particular problem of "red rot" in leather which is caused by attack from sulphur dioxide pollution in the atmosphere. The treatment is carried out dry, the leather being suspended in the fumes above concentrated ammonia solution. Then:

$$2NH_3 + H_2SO_4 \rightarrow (NH_4)_2SO_4$$

You can regard this reaction as the formation of hydrogen bonds through the weakly bound hydrogen atoms on the sulphuric acid molecules to the negative end of the ammonia molecules, though the transfer of H atoms is probably more definite than this suggests

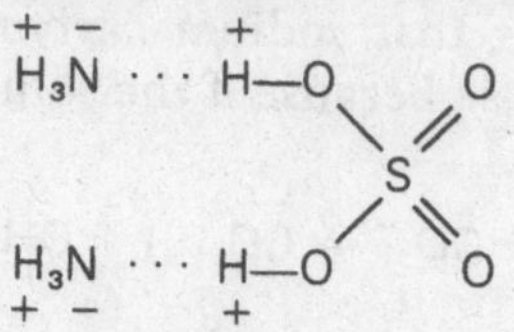

$$\begin{array}{c} \overset{+\ -}{H_3N} \cdots \overset{+}{H}\!-\!O \quad O \\ \quad \diagdown \ // \\ S \\ \quad / \ \diagdown\diagdown \\ \underset{+\ -}{H_3N} \cdots \underset{+}{H}\!-\!O \quad O \end{array}$$

Figure 6.10

because ammonium sulphate forms ionic crystals $(NH_4^+)_2SO_4^{2-}$. Because no water is involved, the ammonium sulphate stays on the leather. Any moisture absorbed from the atmosphere cannot release acid H_3O^+ ions from the sulphuric acid because the hydrogen atoms are already bonded into NH_4^+ ions. This deacidification treatment has seemed to work and has caused no damage over a period of time after treatment. However, there are doubts as to whether ammonium sulphate is going to be stable on leather; under some conditions some ammonia will be lost, reforming sulphuric acid to continue its attack on the collagen (protein) fibres of the leather.

C The measurement and control of pH

C1 The measurement of pH

The colour of many organic substances depends upon hydrogen ion concentration. These can be used as **indicators**, the most common being litmus which is red in acid, and blue in alkaline solution. "Universal Indicator" undergoes subtle changes of colour which can be compared with those on a chart over a wide range of pH and can thus be used to estimate the pH of a water solution.

indicators

A **pH meter** is an instrument for measuring the hydrogen ion concentration, and shows whether a solution in water is acid or alkaline. The probe of the pH meter is designed to be immersed in a water solution. There is a special membrane in the probe through which only H_3O^+ ions can diffuse and the flow of electricity behind this membrane is measured. The electricity can only flow by being transported by the H_3O^+ ions, and the instrument is calibrated to display the flow of current as a measurement of the concentration of these ions in the original solution, that is, its pH.

pH meter

Proper measurements for pH require the temperature to be taken into account. Readings intended for comparison should be carried out at the same temperature, or compensation should be made if they are taken at different temperatures. Standard measurements are taken at 25°C. Your pH meter will have a control which can be adjusted for different temperatures and may also have its own thermometer built in. (If not, a separate thermometer is necessary to read the temperature of the liquid.) Readings taken at different temperatures without measurement of temperature and compensation for that difference are not comparable one with another, and are consequently virtually useless. The simple explanation for the effect of temperature is that, as you will remember, the ionisation (breaking up) reactions in water are accelerated at higher temperatures while the recombination reaction is slowed as ions are less likely to catch each other.

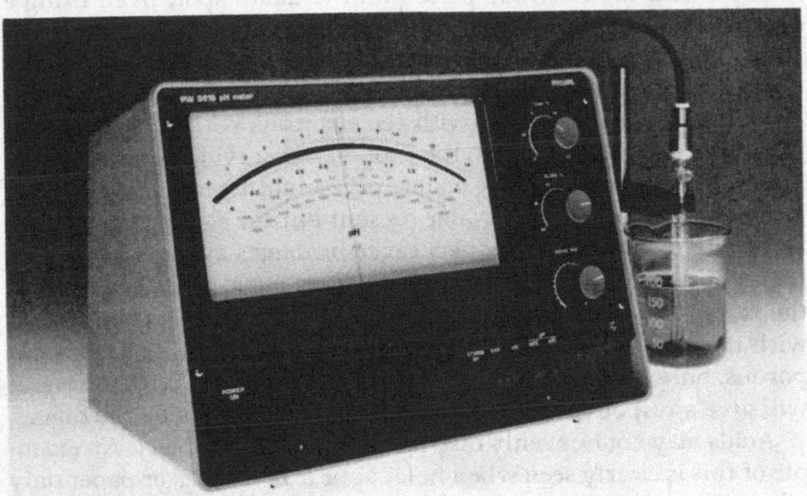

Figure 6.11 *A pH meter.*

pH meters are used to measure the pH of washing solutions before and after rinsing an object, and to indicate the amount of acid present in an object, in situations where a sample can be sacrificed to be macerated (soaked and broken up) in distilled or deionised water. This has to be done because all the acid must be extracted and dissolved in the water for the pH to be registered accurately on the meter. It is important to have the solution thoroughly mixed before taking a sample for pH measurement, because differences may occur throughout the bulk. For example, in washing tanks there may be temperature differences as cold solution sinks and warm rises. Indicator papers are less sensitive than pH meters but both have their uses. If an indication of whether the solution is acid or alkali is all that is needed, to use a pH paper only may well be adequate.

Problems with pH readings

It is important to remember that when we talk about the distinction between acids and alkalies, we are talking about what happens to *water* when these substances are added to it. Thus, the measurement of pH *only has validity when water is present.* A pH value for the solution left after washing an object in an organic solvent, containing no water, has no meaning. Similarly, it is impossible to measure the pH of a dry object, such as a piece of paper. The dry solid may be loaded with an acid salt such as sodium bisulphate which will become actively acid when wetted but the pH of that solution will depend upon its concentration. Thus, even if you "moisten" the solid, the meter can do no more than tell you of the presence of acid – the particular number indicated will have far more to do with the *amount* of water you have used than with any other features of the object. That is why there are conventional procedures for preparing water extracts from solid samples which will give consistent pH readings. Some electrodes for use with pH meters are designed to be used directly on a surface. A drop of water is applied to, say, a piece of paper and the electrode placed on the damp spot. Even using a surface electrode it is difficult to get a useful numerical measurement. The ions present in the material have to transfer themselves to the drop of water that has been put on the surface and, through that water, to make contact with the electrode which measures the pH. If the material is porous, the drop of water will keep disappearing into it and without that water, pH cannot be measured.

surface pH readings

Not only must there be water present but the acidic ions must *go into solution* in it. This may also cause problems as the acid may be there, but locked up in some way, perhaps in an adhesive. Perhaps the water does not adequately wet the surface and get into contact with the ions so they are not dissolved. Again, if the surface is not porous, only the acid sitting on the surface in contact with the water will give a response and there may be much more inside the object.

Acids may not be evenly distributed through an object. An example of this is clearly seen when holes appear in textiles or paper only where iron mordants or iron ink have been. Iron encourages the reaction between sulphur dioxide and moisture to produce sulphuric acid.

Finally, it is quite likely that a surface reading on pH might be attempted on an object that will not stand washing. There is then the risk of forming rings where the water drops have been placed.

You must remember that surface electrodes are intended only to demonstrate the *presence* of acid. The actual numbers do not mean much and quoting them to even one decimal place, let alone two, is a nonsense. Properly used, however, pH meters are valuable tools enabling intrinsically hazardous chemical treatments to be brought under control and performed effectively and safely.

C2 Buffer solutions

There may be occasions when it is important to have control of the pH of a solution, not merely to leave the solution acid or alkaline. Solutions which control their own pH, keeping it constant when more acid or alkali is added, are called **buffer solutions**. You will, no doubt, have made them by dissolving the appropriate manufactured tablets in the prescribed amount of water, without having the scientific information to understand how they work.

buffer solutions

Their action depends upon the kind of reversible reactions described for weak acids and bases. This self-controlling property can be achieved by mixing a weak acid with a soluble salt of that acid. A typical example of an acidic buffer solution is a solution of acetic acid, CH_3COO-H, and sodium acetate, $CH_3COO-Na$. The salt is completely ionised in solution while the acid is only partially ionised. In solution the following reactions occur:

a) $NaAc \rightleftharpoons Na^+ + Ac^-$

b) $H_2O + HAc \rightleftharpoons H_3O^+ + Ac^-$

c) $H_2O + H_2O \rightleftharpoons H_3O^+ + OH^-$

(Ac represents the group CH_3COO)

You can see how the amount of OH^- present from reaction c is controlled by increased H_3O^+ from reaction b, the extra H_3O^+ raising the probability that OH^- will combine with it to form water. Similarly the ionisation of acetic acid is inhibited by the Ac^- ions from the sodium acetate. The chances of the reaction

$$H_3O^+ + Ac^- \rightarrow H\,Ac + H_2O$$

are increased if there are more Ac^- ions for the H_3O^+ ions to meet. A balance is struck when the rates of ionisation and recombination of water are equal and the rates for acetic acid are equal. The exact concentration of H_3O^+ will be determined by how much Ac^- has come into solution from the salt. The equilibrium pH can thus be chosen by deciding on the proportions of acid and salt.

The interesting property of such mixtures is their response to the addition of H_3O^+ or OH^-. Consider them in turn:

- If H_3O^+ is added, these ions are likely to meet Ac^- and recombine to form HAc.
- If OH^- is added there are many H_3O^+ ions for them to recombine with. The resulting fall in H_3O^+ concentration reduces the rate of recombination of H_3O^+ with Ac^- so the forward reaction

$$HAc + H_2O \rightarrow H_3O^+ + Ac^-$$

wins for a little while until the H_3O^+ population is re-established.

Thus any attempts either to acidify or to make the solution more alkaline are compensated, and the pH remains constant. Of course there are limits to the amount of additional acid or alkali which the acetate reactions can handle.

The sodium acetate/acetic acid system will buffer solutions in the range pH 3.5 to 5.5. There are also systems which will buffer alkaline solutions, (for example, sodium hydroxide and disodium hydrogen phosphate) for pH 11 – 12.

One use of buffer solutions is to calibrate pH meters; the known pH of the standard solutions is not affected by impurities. The fact that buffer solutions maintain a constant pH when acid or alkali is added is useful in conservation, for keeping a process (electrolysis, washing, or bleaching) at the most effective pH for treatment or for the safety of the object (for example, not too acid for cellulose or too alkali for protein fibres).

buffer

Sometimes the term **buffer** is used (quite differently) to describe a substance which is used to mop up acid coming from the atmosphere and hence to prevent damage to the object. Calcium and magnesium carbonates in paper are examples of this. They can be applied to paper to protect it from acid (atmospheric pollution).

$$\underset{\text{pollution}}{H_2SO_4} + \underset{\text{"buffer"}}{CaCO_3} \rightarrow \underset{\text{neutral salt}}{CaSO_4} + H_2O + CO_2.$$

The same substances can, of course, be used to deacidify already acid paper. If calcium carbonate is used to mop up acid from the atmosphere it is not acting in quite the same manner as the buffer systems just described. It is more accurate to call it an **alkali reserve.**

alkali reserve

7

Cleaning by chemical reaction

A Energy changes in reactions

B Oxidation and reduction reactions

B1 *Oxidation reactions in cleaning*

B2 *Reduction reactions in conservation*

B3 *The direct use of hydrogen or oxygen*

C Sequestering compounds

D Catalysts

D1 *Enzymes – Nature's catalysts*

Cleaning by chemical reaction

Except for the one instance of dissolving ionic bonded crystals everything in this book has been about breaking *secondary* bonds between dirt and object to release the one from the other. When a method of doing that cannot be found it is time to attack the *primary* bonds holding the atoms together in the dirt molecules. Book 1 showed you how those primary bonds are particular arrangements of electrons between atoms and explained how chemical reactions involve breaking existing bonds to allow new and different ones to be made. The problem of cleaning using chemical reactions is to find *reagents* which will attack the dirt without touching the object. This ideal is sometimes possible but more often we have to be satisfied with attacking the dirt more rapidly than the object, so that damage is kept to a low level.

A Energy changes in reactions

You know that many chemical reactions generate heat. For example the burning of methane in air is used all the time, primarily to produce heat, the chemical products (carbon dioxide and water) being unimportant in that context.

$$CH_4 + 2O_2 \rightarrow CO_2 + 2H_2O + \text{heat}$$

A chemical used in this way is called a *fuel* but production of heat in reactions is not limited to fuels. When plaster of Paris sets, heat is produced. Resins used for mounting samples for microscopy and resins used to consolidate friable wood and stone liberate large amounts of heat as they solidify and this can cause damage to the object or the sample.

A reaction which leads to the production of heat is called **exothermic**. The appearance of heat suggests an increase in the kinetic energy of the molecules. Since no energy is being put into the

exothermic reactions

reaction from outside, the increase in kinetic energy must come from a decrease in potential energy. The molecules rearrange themselves into a more favourable state which has a lower potential energy. The products of the reaction are more stable than the starting materials (reactants) and the difference between these two energy states is the observed increase in kinetic energy – heat.

When plaster of Paris and water are mixed together in the right proportions there is nothing that can be done to stop the mixture setting. The reaction is spontaneous; it wants to happen, there is a desire to reach the more stable state.

system This can be interpreted as a *decrease* in the free energy of the system (in this context, the word **system** means the particular thing you are looking at). We can use an equation like the one we used in Chapter 3 to explain this.

$$\Delta G = \Delta H - T\Delta S$$

If heat is formed during the reaction, ΔH will be negative; the heat is *lost* from the system. As the plaster sets, the once-mobile water becomes locked into a solid substance. The disorder of the system has decreased and so the *entropy* change, ΔS, is negative. ΔG will be negative if TΔS is smaller than ΔH. As we can observe that the setting reaction is spontaneous, we know that this must be the case.

The larger the negative value of ΔG, the greater is the desire for the reaction to take place. If there is an increase in entropy when an exothermic reaction occurs, both the heat and the entropy parts of the equation will be working to make ΔG large and negative. An example of this is in the burning of fuels. The liberated heat indicates a move to a more stable potential energy state, that is, less total energy in the bonds. ΔH is negative. In the burning of a hydrocarbon such as hexane:

$$2C_6H_{14} + 19O_2 \rightarrow 12CO_2 + 14H_2O$$

there is an increase in the total number of gaseous particles; 21 particles react to form 26. There is an increase in disorder; ΔS is positive.

Not all reactions liberate heat. There are some reactions where heat is spontaneously absorbed. The most commonly observed are physical reactions. Acetone evaporates spontaneously at room temperature; the remaining solvent becomes chilled and water may be condensed from the atmosphere onto the container. The secondary bonds between the acetone molecules are being broken as isolated molecules of vapour are formed. This means there is an increase in inter-molecular potential energy; ΔH is positive. However, the disorder on going from liquid to vapour is increased, so that entropy has increased and ΔS is positive. Since you have observed that the evaporation of acetone at room temperature is spontaneous you know that ΔG must be negative, so TΔS will get even larger if the temperature T is increased. This means that the evaporation of acetone will be more favoured at high temperatures, which you know to be true. What you *have* learned, though, is that reactions where there is an increase in entropy are more likely to occur as the temperature increases.

endothermic reactions

Reactions that absorb heat are called **endothermic**. There are few examples of endothermic chemical reactions in a conservator's everyday experience, but the smelting of metal ores is one example of an endothermic reaction that produces an unstable material.

Although the intermediate reactions are complicated, the overall effect of smelting ores is to start with the ore plus coke and to end with the metal plus carbon dioxide.

$$\underset{\text{iron oxide}}{2Fe_2O_3} + \underset{\text{coke}}{3C} \rightarrow \underset{\text{iron}}{4Fe} + \underset{\text{carbon dioxide}}{3CO_2}$$

Five particles have reacted to become seven, three of which are gaseous. At a high enough temperature the entropy consideration takes over ($T\Delta S$ becomes large enough to outweigh the positive ΔH) and the reaction proceeds.

The roasting of limestone to form quicklime (see Book 1, Chapter 3) is another example of the entropy factor dominating at high temperatures.

At room temperature the reactions described are not spontaneous as ΔG is positive. But then, if you look at the reaction going the other way, ΔG is negative. At room temperature there is a spontaneous tendency for iron to turn to rust and for lime to react with carbon dioxide to form calcium carbonate.

Many deterioration reactions, such as the degradation of polymers, result in an increase in the number of molecules and in the mobility of the molecules produced. Thus the general tendency towards greater entropy in the world means that the highly ordered objects people make and use become slowly more disordered. Conservation is largely an attempt to come to terms with entropy.

How chemical reactions occur – another look

In Book 1 you learned to think of a chemical reaction as the result of collisions between molecules. These violent collisions cause the fragmentation and rearrangement of the molecules. If on reaction there is a drop in free energy then there is no tendency during further collisions for the newly formed products to react again, to re-form the initial reactants.

We have seen that when a chemical reaction occurs there is a decrease in the free energy of the system as it goes from reactants to products. This is shown in Figure 7.1.

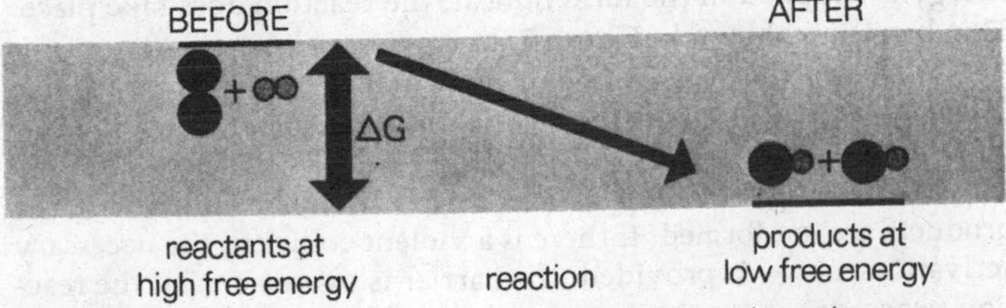

Figure 7.1 *Changes in free energy during a chemical reaction.*

If ΔG is large and negative the reactants *want* to become products. The reactants are said to be **thermodynamically unstable** with respect to the products. Yet not all possible reactions for which there would be a large drop in free energy actually take place. Iron artefacts are unstable in air but many have lasted for hundreds of years. Cellulose is thermodynamically unstable in air. The heat of combustion and the entropy drive on going from one large immobile molecule to many gaseous water and carbon dioxide molecules, suggest that there would be a tremendous decrease in free energy on degradation. Yet objects of wood, paper and cotton survive in air at room temperature, so why do reactions *not* occur which on free energy grounds *ought* to happen?

thermodynamic instability

For a reaction to take place a decrease in free energy is essential. But there must also be collisions. The greater the number of collisions the faster the reaction will occur. The number of collisions depends on how fast the molecules are moving; how crowded together they are; how large the molecules are and how well mixed the reactants are. Just from the idea of reactions being caused by collisions you can see that:

- An increase in temperature will cause reactions to go faster.
- The reaction of two solids will be much slower than the reaction of solutions of those solids.
- Concentrated solutions react more rapidly than dilute ones.
- A divided solid will dissolve more quickly than a large lump.

The amount of thermodynamic instability is the same in all these cases, as this is determined only by the reactants and the products. The **rate of reaction** is different in each case and is determined by factors other than the desire to reach a more stable state.

rate of reaction

Temperature dependence of reactions

The degradation of many historic objects is the result of gases (oxygen, water vapour) reacting with solids (cellulose, copper, iron). The reaction is limited to the surface of the solid so the number of collisions is limited but even so the reactions are very slow. Also, at ordinary temperatures, the collisions that are taking place may not be violent enough to break any existing bonds – a preliminary step in the reaction. With a piece of wood in air, the concentration of cellulose and oxygen is high enough to support combustion; yet it does not take place. There is some sort of barrier to reaction. But if energy is supplied in the form of heat, the reaction does take place. This barrier is shown in Figure 7.2.

The energy that has to be put in to make a reaction possible is called its **activation energy**. Before collision the reactants have a certain amount of energy. If a collision occurs with less than the activation energy the top of the barrier is not reached and the products are not formed. If there is a violent collision, the necessary activation energy is provided, the barrier is overcome and the reaction proceeds.

activation energy

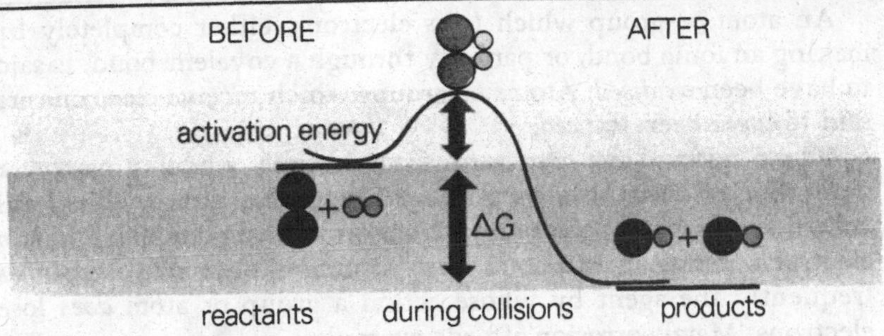

Figure 7.2 *Energy changes during a reaction, shown as a curved line.*

The number of molecules which have enough kinetic energy to make suitably violent collisions is very small indeed. This number increases as the temperature rises, as you might expect, but it is not a simple relationship. The number of sufficiently energetic molecules increases *exponentially* as the temperature rises. For typical chemical processes that occur at room temperatures, only one molecule in 10^{18} has enough energy to cause a collision that will result in reaction. The number is doubled for a 10°C rise but if the temperature is raised by 100°C the number of sufficiently energetic molecules increases *a hundred thousand* times.

With very exothermic reactions (such as combustion) the kinetic energy imparted to the reacting molecules is transferred to other molecules and provides the activation energy for further reaction. The reaction continues, often violently. However, for many reactions the necessary kinetic energy must be provided by keeping the temperature high, so that there is a sufficient number of energetic molecules which can collide successfully.

The activation energy is what determines whether a reaction goes quickly or slowly. If the activation energy is high the reaction may not take place appreciably at room temperature even if there is a net gain in stability to be made by reacting.

Just as entropy plays a part in determining whether substances ought to mix or whether chemicals ought to react; it is also involved in determining the rate at which a reaction takes place.

A certain degree of order may be necessary for a sufficiently energetic collision to be successful. Polymerization, for example, requires a collision at a particular reaction site, at the end of a long chain molecule. No number of collisions at other points on the chain will lead to reaction. Reactions which require a particular orientation of the colliding molecules will usually have high activation energies because of the entropy factor.

B Oxidation and reduction reactions

Whenever a chemical reaction occurs there is a regrouping of atoms, and so there must also be changes in the pattern of electron distribution for each atom or group.

An atom or group which *loses* electrons, either completely by making an ionic bond, or partially through a covalent bond, is said to have been *oxidised*. Atoms or groups which receive electrons are said to have been *reduced*.

These terms have persisted from a time when a narrower definition of them was current, although the term *oxidised* still makes some sense because oxygen shows a strong tendency to gain electrons. Since it is also a very common element, oxygen is frequently the agent by whose action a group or atom *does* lose electrons. Metal corrosion affords examples:

$$2Fe + O_2 \rightarrow 2Fe^{2+}O^{2-} \quad \text{ferrous oxide}$$

$$4FeO + O_2 \rightarrow 2Fe^{3+}{}_2O^{2-}{}_3 \quad \text{ferric oxide}$$

In the first reaction each iron atom *loses* two electrons to form *ferrous* oxide. Further attack by oxygen causes each iron atom to lose another electron to form *ferric* oxide.

oxidation Thus, a change from metal to metal oxide is **oxidation** and so is a change from an -ous to an -ic condition.

However, oxidation need not necessarily involve oxygen. The black tarnish on silver is silver sulphide. We say that the silver has been oxidised because it has lost electrons – but it has been *oxidised* by sulphur.

reduction If the word **reduction** seems a strange opposite to *oxidation* it may help to imagine the chemists of the eighteenth century weighing constituents of chemical reactions. *Reduction* was an obvious word for what happened when a reaction reduced the weight of a compound. Mercuric oxide loses weight when roasted; tinstone and charcoal mixed and heated also lose weight. As equations you can see what happens:

$$2HgO \xrightarrow{\text{heat}} 2Hg + O_2$$

$$SnO_2 + C \xrightarrow{\text{heat}} Sn + CO_2$$

In both cases the loss of weight is due to a gas being formed and leaving the reaction site (but the early chemists did not realise this). You can see that in both the oxygen has given back the electrons which it had taken from the metal atoms. The oxides have thus been *reduced* to the metals.

The example just given should have made the technical importance of oxidation and reduction clear. These are the reactions through which metals corrode (oxidation) or are won from their ores (reduction).

B1 Oxidation reactions in cleaning

Oxidation (taking electrons out of a bond) is one way of disrupting covalent compounds. Many of the materials of which objects are made and much of the dirt you want to remove are complex covalently bonded organic compounds. You may find that particular unwanted parts can be attacked in ways which involve critical bonds changing from an equally shared covalency to more polar forms, producing at least a partial loss of electrons from some atoms – oxidation.

bleaching

One use of oxidation in conservation is **bleaching**, for which several chemicals are used. All of these release oxygen *atoms*, O, an even more reactive form than oxygen *molecules*, O_2. The molecule, O_2, is, of course, the stable form where the two atoms share electrons to give each a full outer shell. Oxygen in the atomic form, O, is unstable and can only exist for a very short time before reacting with some other substance. The simplest and most natural source of single oxygen atoms is atmospheric oxygen in the presence of sunlight. Energy in the sunlight disrupts a minute proportion of the oxygen molecules $O_2 \rightarrow 2O$, thus producing some oxygen in the active atomic condition. This is partly why sunlight fades dyestuffs in curtains and why light levels on organic materials should not be intense. Exposure to sunlight was the traditional method of bleaching linen cloth and has recently been recommended as a method for bleaching yellowed paper without introducing any chemicals such as chlorine, which might have harmful long-term effects.

Other reagents yielding oxygen atoms are sodium hypochlorite (domestic bleach), hydrogen peroxide and chloramine-T which react as follows:

$NaOCl \rightarrow NaCl + O$ (remember that O only has a transitory existence)

$H_2O_2 \rightarrow H_2O + O$

$$CH_3\text{–}C_6H_4\text{–}SO_2-N^-ClNa^+ + H_2O \rightarrow CH_3\text{–}C_6H_4\text{–}SO_2\bar{N}HNa^+ + HCl + O.$$

Of these, peroxide is the most widely used in conservation because the only residue is water which is usually harmless and dries off anyway. This is certainly not the case with the others which will leave reaction products (such as common salt) and solid unused bleach on the object when it is dried. Long-term attack on textile and paper fibres by unused bleach should not be risked. For a while it was thought that chloramine-T residues were harmless, but now it is considered advisable to wash them out too, using either water or alcohol.

What is a stain?

Before we can suggest how bleaches remove stains we must say something about what types of compounds are coloured and which are colourless. The difference between a dye and a stain is that "difference" between a weed and a flower mentioned at the beginning of the book. The two may be chemically similar but one is desirable and the other is not. Moreover, even if a stain molecule can be converted into another related substance that is colourless, the matter may still remain when its immediately disagreeable quality has been removed.

What is colour?

You are familiar with the idea that radio signals are transmitted as waves of some sort. The words *wavelength* and *frequency* (the number of waves per second) are in everyday use when talking about tuning radios. You may also know that light is transmitted as **electromagnetic waves**, similar to radio waves but of much shorter wavelength and much greater frequency. The human eye can detect these waves within a narrow band of wavelengths, and the brain interprets different wavelengths as different colours.

electromagnetic waves

Light from the sun contains a mixture of all the visible wavelengths. When light falls on an object some light is absorbed and some reflected. If all the wavelengths are equally reflected the object appears white, but if they are completely absorbed the object appears black. If the wavelengths which the brain interprets as blue are absorbed then the object appears yellow. That is, if some wavelengths are preferentially absorbed by the object, the eye and brain interpret the remainder, which are reflected as having a particular colour.

But why do some objects (that is to say, some *compounds*) absorb particular wavelengths of light? Light is a form of energy and the shorter the wavelength the greater is the energy of the light. It is well known that the short wavelength ultraviolet light is very energetic, that is, very destructive. When a molecule absorbs light it absorbs energy – it goes to a state of higher energy than it had before.

In Book 1 you learned that a molecule could be considered as numbers of electrons in molecular and atomic orbitals surrounding the nuclei of the component atoms. When a molecule absorbs light energy an electron is moved from an occupied molecular orbital to one that is not normally occupied.

In any compound the difference in energy between the normal state and the *excited* (more energetic) state is fixed, because it is determined by the types of atom present and how they are held together.

Molecules with similar structures will have similar types of molecular orbitals. In similar molecules, the difference in energy between the normal bonded structure and that after light has been absorbed will be about the same. In some structures this energy difference will be equal to the energy of some of the light waves that are visible to the human eye. The reflected light will then be deficient in some wavelengths, so these compounds will appear coloured. More frequently the energy difference corresponds only to the energy of ultraviolet light and so, as no *visible* light is absorbed, the compound will appear white or colourless.

The types of molecular orbital that are associated with absorption of light are those that are related to the presence of double bonds and benzene rings. On their own such structures absorb only ultraviolet light but if the double bonds or aromatic rings are **conjugated** (that is, joined in long chains so that double bonds alternate with single bonds) the compounds are coloured.

conjugated bonds

The naturally occurring substance ß-carotene is often used as a food colourant. Its molecular structure consisting of a long chain of conjugated double bonds.

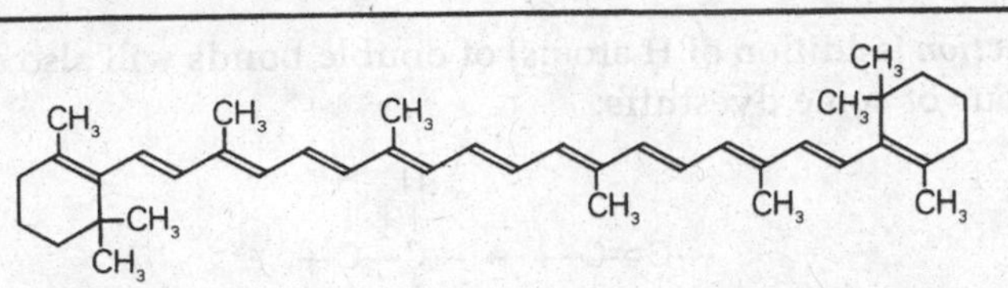

Alizarin madder has benzene rings conjugated to C = O double bonds.

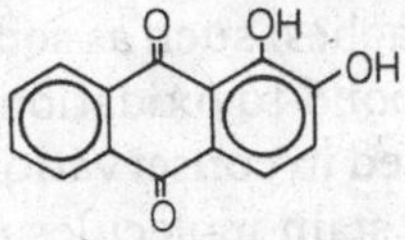

Indigo has benzene rings conjugated to both C = O and C = C double bonds.

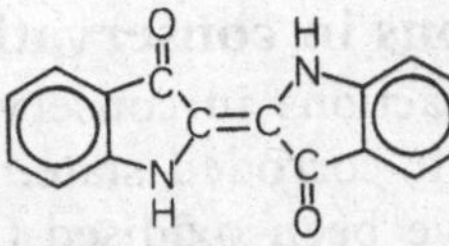

Congo red, another dye, has several aromatic rings conjugated by N = N double bonds.

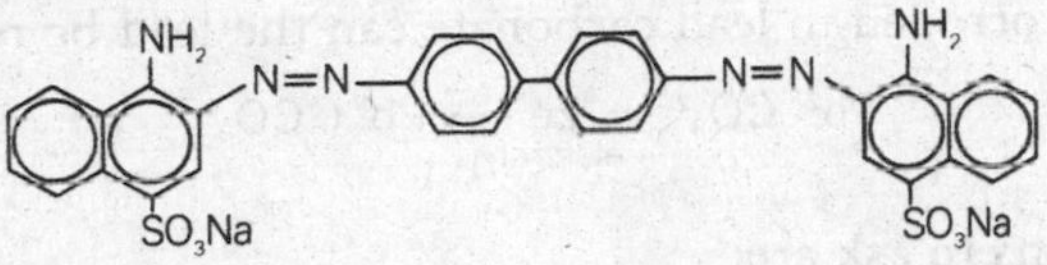

The groups such as – C = O, – C = C – and – N = N – which confer colour on molecules are known as **chromophores**.

chromophores

How bleaches work

If one of the double bonds in the conjugated chain can be attacked by oxygen then the conjugation can be broken and the compound will no longer be coloured. When you remove a stain using bleach you are oxidising the chemicals which constitute the stain. Since the colour-giving parts of molecules are those rich in electrons, they are susceptible to oxidation, that is, to loss of electrons. The common colour centres, which consist of alternate single and double bonds along a chain of carbon atoms, are changed sufficiently to become colourless if just one double bond yields electrons to an attacking oxygen atom. The following kind of reaction may occur:

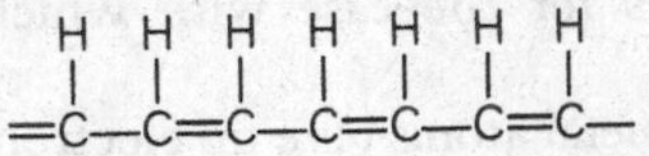

becomes:

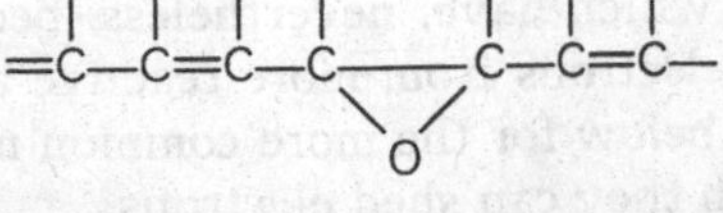

The mutual sharing of electrons along the chain is disrupted and the colour disappears.

Reduction (addition of H atoms) of double bonds will also remove the colour of some dyestuffs:

$$\begin{array}{c} -\underset{\mathrm{H}}{\overset{|}{\mathrm{C}}}=\underset{\mathrm{H}}{\overset{|}{\mathrm{C}}}- \end{array} \rightarrow \begin{array}{c} \mathrm{H}\ \ \mathrm{H} \\ -\overset{|}{\underset{|}{\mathrm{C}}}-\overset{|}{\underset{|}{\mathrm{C}}}- \\ \mathrm{H}\ \ \mathrm{H} \end{array}$$

and so there are both oxidation bleaches, such as hydrogen peroxide (H_2O_2), and reduction bleaches, such as sodium dithionite ($Na_2S_2O_4$). More dyes and stains respond to oxidation than to reduction and so oxidation is frequently used in conservation. The similarity between the structures of dye and stain molecules means that bleaches must be used carefully to avoid damage.

B2 Reduction reactions in conservation

One use of reduction reactions in conservation is the attempt to recover metals from their corroded state. When a metal forms a compound its atoms have been oxidised (lost electrons), forming positive ions. Reduction (putting electrons back into these ions) would give, as one product of reaction, the metal itself. For example, if lead has corroded to lead carbonate can the lead be regained?

$$Pb^{2+}CO_3^{2-} + \underset{\text{electrons}}{2e^-} \rightarrow Pb + CO_3^{2-}$$

The questions to ask are:

- How can the necessary electrons be provided?
- Where will the carbonate ions go?
- How much damage will there be to the uncorroded metal beneath the corrosion layer?
- How much damage is acceptable?
- What will the metal look like when it is recovered?

Electrochemical reduction

In Book 1 you learnt that metals are held together in bulk by a special primary bond in which some electrons are shared by *all* the atoms. These electrons are considered to be free to move around in the metal, which accounts for the ease with which metals conduct electricity.

Remembering that metal atoms give up electrons readily in forming ionic compounds it can be deduced that the metals most ready to lose electrons would be the best source. Metal atoms less able to shed electrons but which have, nevertheless, become ions, can be induced to accept electrons from more reactive atoms. Metals can thus be ranked (as below for the more common metals) in order of the ease with which they can shed electrons.

The most reactive metals corrode too quickly in the air or water for them to make useful materials for most objects (although

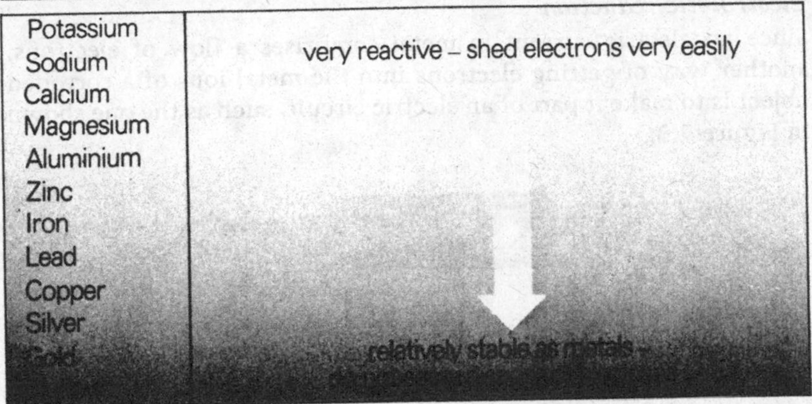

aluminium forms an oxide which adheres tightly to the metal and protects it from further corrosion). The corrosion products on an artefact made of any metal low in the list given above can in theory be reconverted to metal, if intimate contact can be made with an element higher up the list. The *more* reactive metal becomes *ionised,* giving electrons to the ions of the *less* reactive element which thus becomes metal again.

Demonstration

Put a clean iron nail into a solution of copper sulphate. You will see red copper form on the surface of the nail. Eventually the nail will dissolve.

A conservation example is provided by one method of cleaning silver. Aluminium foil is packed around the silver object and put into an alkaline solution such as sodium carbonate in water. The alkali strips the surface coating of oxide from the aluminium and the electron exchange can then occur:

$$\underset{\text{metal}}{Al} + \underset{\text{ions}}{3Ag^{+}} \rightarrow \underset{\text{ions}}{Al^{3+}} + \underset{\text{metal}}{3Ag}$$

electrochemical reduction

Electrochemical reduction has also been used to reduce the corrosion on bronzes, using zinc granules as the other metal. Simple though such methods are from a chemical point of view, they are not without hazard for conservation use. If the object is heavily corroded the treatment may be too drastic in its effects, and even with less-corroded objects the effect may be uneven. This may be due to poor contact between the two metals and the solution and, consequently, insufficient movement of the solution to provide uniform distribution of the ions taking part in the reaction. *Electrolytic reduction* (see next section) has proved to be more controllable, and it is easier to see what is happening to the object using this method than with electrochemical reduction (when it is hidden by a heap of zinc granules or aluminium foil). Thus, it is obviously easier to stop the treatment whenever necessary.

Electrolytic reduction

Since an electric current in metal comprises a flow of electrons, another way of getting electrons into the metal ions of a corroded object is to make it part of an electric circuit, such as the one shown in Figure 7.3.

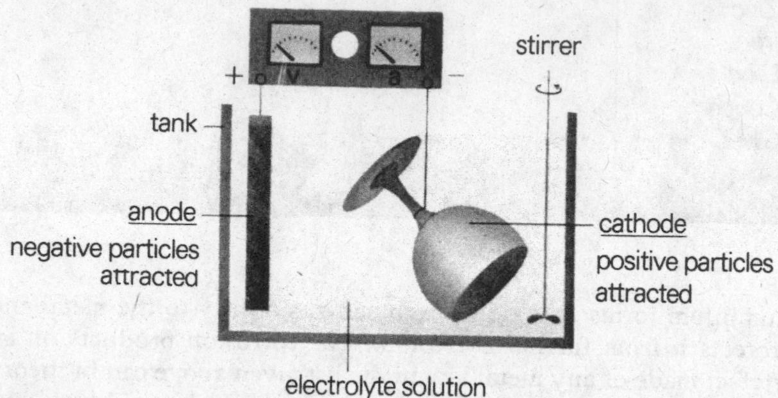

Figure 7.3 *Circuit for electrolytic reduction.*

electrolytic reduction

The circuit consists of a battery (or other direct current supply), an on-off switch and means for controlling and measuring the electric current flowing (variable resistance and ammeter respectively).

The battery can be seen as an electron pump which pushes electrons out of the terminal marked negative where the object is connected. The current flowing away from the object must be carried by some particles other than electrons as these are consumed in the reaction of ion to metal atom. The circuit continues through an ionic solution (called the **electrolyte**) and it is *negative ions* which carry the current through the liquid. When the ions reach the **anode** they are oxidised (giving up their extra electrons to the metal plate) and electrons continue the flow of current through the circuit.

electrolyte
anode

The chemistry of the process is best looked at using a specific example. The method has been used to reduce lead carbonate corrosion on lead artefacts. The lead is made the **cathode** of the circuit (that is, the part that receives electrons from the battery). The intended reaction will take place at the surface of the object, in the corrosion. What is intended is:

cathode

Pb^{2+}	+	$2e^-$	→	Pb
lead ions in the corrosion		electrons from the battery		lead metal

The lead atoms which have been made from the lead ions can no longer hold on to the carbonate ions, so these drift away into the liquid. The rest of what happens is not really important from the conservation angle because the reduction of Pb^{2+} ions to Pb atoms has already been achieved. However, that cannot be the only effect because the electrons must be returned to the battery. For this reason the solution is made alkaline and at the anode OH^- ions give up electrons to the wire part of the circuit:

$$4OH^- \rightarrow 2H_2O + \overset{\text{gas bubbles}}{O_2\uparrow} + 4e^-$$

Unless conditions are carefully controlled the reduction only results in a non-cohesive mass of metal which falls off leaving a cleaned object. But if sufficient control is possible the reduced corrosion products will build up as a cohesive layer of metal and something like the original form of the surface may be created. Attempts to convert corrosion products back into metal while retaining the object's form are called **consolidative reduction**.

consolidative reductio

Electrolytic reductions have been suggested for several conservation problems but are not without difficulties. The object must be a conductor of electricity, that is, a metal and electrical contact must be made to it which may involve damage. The corrosion layer it is hoped to reduce is an electrical *insulator* so passage of current through it is likely to be difficult or impossible. Moreover, if a reaction can be started it will begin at the metal/corrosion interface, perhaps deep in the corroded layer where you cannot see what quality of metal is being made. Also, changes in volume due to the reaction could cause outer layers of corrosion to flake off.

What is *supposed* to happen and what *actually* happens in electrolytic treatments have an uncertain connection. Many variables in the object, in the electrolyte (composition, concentration, pH, temperature) and in the current chosen can influence the results. It is important that specifications for electrolytic reduction processes be followed rigorously and even then, because objects themselves are so variable, the progress of each treatment must be monitored.

Chemical reducing agents

Where electrochemical and electrolytic methods are inappropriate (for example when the object is not metal), chemical reactions in which spare electrons become available can be used to effect reduction. Substances having this property are called **reducing agents**, and common use is made of those in which sulphur donates electrons to oxygen to make negative ions. In Book I you met the sulph*ite* ion, SO_3^{2-}, and the sulph*ate* ion, SO_4^{2-}. There are several others:

reducing agents

ion	name	O to S ratio
$S_2O_3^{2-}$	thiosulphate	4
$S_2O_4^{2-}$	dithionite	3
HSO_3^-	bisulphite	2
HSO_4^-	bisulphate	$1\frac{1}{2}$

Ions with less than 4 oxygen atoms per sulphur atom can be *oxidised* to sulphate or bisulphate thus allowing something else to be *reduced*. In alkaline solution such reactions as these occur:

$$HSO_3^- + 2OH^- \rightarrow HSO_4^- + H_2O + 2e^-$$

$$S_2O_4^{2-} + 8OH^- \rightarrow 2SO_4^{2-} + 4H_2O + 6e^-$$

These equations must not be taken literally because the electrons would not float around on their own. Several unstable species of ion exist momentarily as the reaction proceeds and it is these that donate surplus electrons, effecting reduction.

Reducing agents can be used to remove mineral stains, notably rust marks. Oxidative bleaches would not be any use on these because the metal atoms are already in an oxidised state. Rust, hydrated iron oxide, can be brought into soluble form by the reducing action of sodium bisulphite allowing stains to be washed away. (This is *not* a suitable treatment for rust stains on paper because bisulphite residues, if they are not precisely neutralised, will become acidic and attack the paper.)

B3 The direct use of hydrogen or oxygen

hydrogen reduction

The direct use of hydrogen as a reducing agent is seen in the **hydrogen reduction** furnace for the removal of chlorides from marine iron artefacts. The main chloride-containing component of iron corroded by seawater has been reported as ferric oxychloride, FeOCl. This can be decomposed by heat alone

$$2FeOCl \xrightarrow{heat} 2FeO + \overset{gas}{Cl_2}$$

but the production of chlorine gas deep within the corrosion layer causes the object to split. Hydrogen introduced into the furnace allows alternative reactions:

$$2FeOCl + H_2 \rightarrow 2FeO + 2HCl$$
$$FeO + H_2 \rightarrow Fe + H_2O$$

Both these reactions are reductions because the iron gains electrons. In the first, ferric ions, Fe^{3+}, become ferrous ions, Fe^{2+}, by the addition of one electron. In the second, two more added electrons bring ferrous ions back to the uncharged metallic atoms.

The first reaction is the important one as it is this step which removes the harmful chloride ions. The extent to which the second reaction actually happens, bringing back the original metal, is uncertain, but, with the chlorides removed, a chemically stable system has been produced. Of course, even if the corrosion products of iron *are* reduced back the metal, its physical texture, shape and metallurgical structure will not resemble that of the original object, because iron corrosion products have such a swollen structure that the original shape has been lost. Further attention will also be necessary to ensure subsequent chemical and physical stability.

Gas plasma methods

The hot and/or wet treatments outlined in preceding sections may often be justly regarded as hazardous for objects, so mention should be made of the possibilities of treatment by gaseous oxidising or reducing agents. Both oxygen and hydrogen are gases but under ordinary conditions react only slowly with objects or their dirt. An alternative approach has received attention from research workers during recent years – the gas plasma method.

gas plasma

A plasma in this context means ionised gas. A common instance is the gas found in fluorescent lamp tubes or sodium vapour street-lights. Once molecules of gas have been ionised they can be accelerated to high speeds in an electric field and made to bombard a surface. Moreover, ionised gas molecules are chemically unstable and therefore highly reactive. Given suitable apparatus (Figure 7.4) a stream of such molecules can be brought to bear on an object, and both oxidation and reduction reactions conducted with great delicacy. Hydrogen gas is used for reduction and oxygen for oxidation. Pure plasmas of these elements would be too fierce, chemically, and so only a small percentage of the reactive gas is used, the rest being the chemically inert gas argon.

The apparatus built so far has restricted the size of objects which can be treated by this method and requires an electrical connection to the object. It has removed the tarnish from intricate jewellery surfaces but has not been so successful on metal threads in textiles.

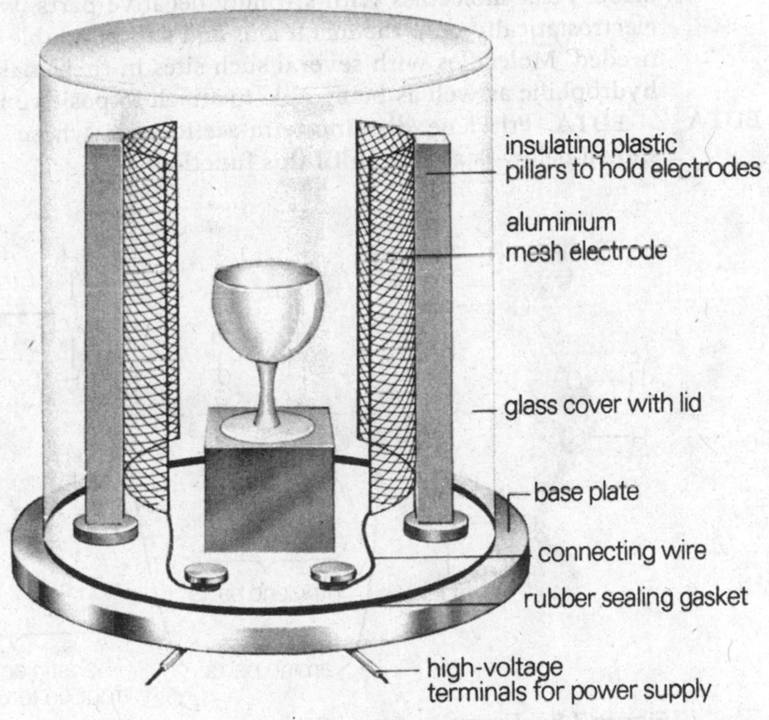

Figure 7.4 *Apparatus for gas plasma treatments. One of several electrode arrangements for generation of a plasma.*

C Sequestering compounds

Chapter 5 showed how water is able to dissolve ionic compounds, that is to attack the *primary* bonds within ionic crystals. This is because the low potential energy state of the ions in the crystal can be substituted by a condition of almost equally low potential energy when ions are surrounded by polar water molecules. The way water molecules can cluster round positive or negative ions to form *hydrated ions* is just one example of many **complexes** between ions and polar molecules. The general technique of forming such complexes is called **sequestering** (meaning separating or cutting off). Water itself is a sequestering agent and so (remembering *miscelles* mentioned in Chapter 5) are detergents, although the term is usually reserved for all instances *other* than these.

complexes

sequestering agents

The blue colour of a solution of copper sulphate originates in the hydration of the Cu^{2+} ion to form a complex ion, $Cu(H_2O)_6{}^{2+}$. If ammonia is used to clean copper or its alloys (brass and bronze) a much deeper blue colouration is sometimes seen. This colour is due to a complex ion, $Cu(NH_3)_6{}^{2+}$, known as the cuprammonium ion. Both complex ions are soluble in water, whereas copper alone is not. By forming a soluble complex from an insoluble compound, it is possible to remove unwanted material, washing it away with water. However, there may be a risk to the material of which the object is made. Polar molecules with strongly negative parts, which can tie electrostatically onto the metal ions and still be souble in water are needed. Molecules with several such sites in their makeup will be hydrophilic as well as being able to attach to positive ions.

EDTA

EDTA, ethylene diamino-tetra-acetic-acid, whose structure is shown below is able to fulfil this function:

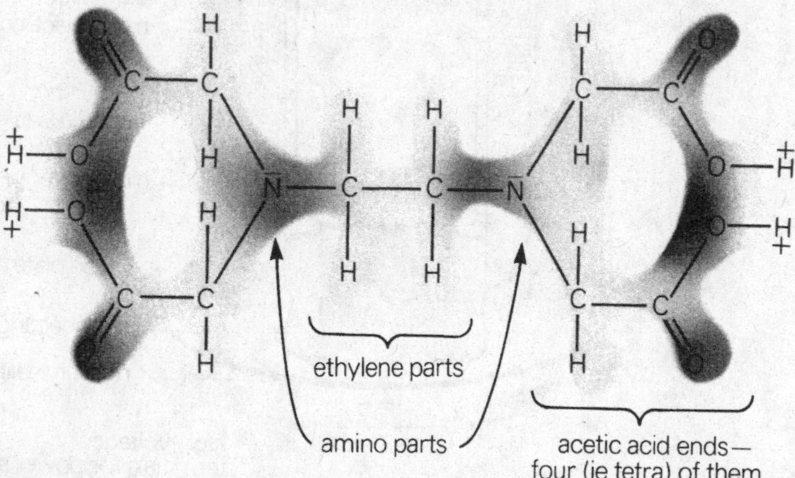

Figure 7.5 *The structure of EDTA (negative sites are shown by the heavily shaded areas).*

The acidic hydrogens (marked +) are often substituted by Na^+ ions in use. The rest of the + charge is distributed thinly over the CH_2 groups. There are now several possible sites of electrostatic attraction for the positive metal ions and each ion can be held by more than one site. When this is possible the substance is called a **chelating agent** (pronounced with a hard 'c' from the Greek for a claw), and the complex ion formed is called a **chelate**. The remaining polar groups ensure the complex is soluble in water through hydrogen bonding from water molecules.

chelating agents

chelate

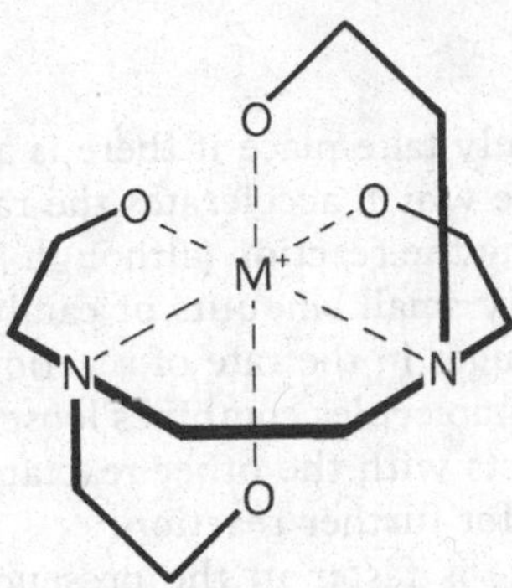

Figure 7.6 *Metal ion surrounded by all six bonding sites in EDTA.*

EDTA is a "weak" acid. That means, you will remember, that only a small proportion of its molecules are ionised when it is dissolved in water. The chelating action is rather more restricted if the acid hydrogen atoms remain in place and so, by changing the pH of the solution, the strength of the chelating effect can be varied. In acid solutions H^+ ions bond on to the $-COO^-$ ends, while in alkaline environments these groups are available to hold metal atoms. The tri-sodium salt (in which three of the H atoms are replaced by ionically bonded Na^+ ions) dissolves at 1% concentration to give a solution of pH9.3 and is a potent sequestering agent. The tetrasodium salt at 10g/l concentration has pH11.3 and bonds even more strongly to positive ions.

By adjusting the pH, some control over the action can be achieved. In use you have to consider the conditions which enable the chelating agent to attack that which you wish to remove without damaging the object. Thus the tetrasodium salt will take iron stains out of limestone or marble but will also take calcium ions from the stone, etching the surface. A lower pH must be used. However, the tetrasodium salt *is* used for removing calcareous deposits from pottery, where the risk of damage to the object is less.

The use of all chelating or sequestering agents carries the risk of their attacking the object underlying the deposit to be removed. Ions may be stripped from glass or ceramics and metals dissolved. Sometimes the proportions of metals in alloys on the surface of metal objects can be upset by dissolving one metal faster than another.

Not all sequestering agents are chelating agents, because not all of them offer the possibility of holding ions between two sites. Preparations such as Calgon (sodium hexametaphosphate, $Na_6(PO_3)_6$) form complexes with Ca^{2+} and Mg^{2+} ions without the chelation mechanism. Calgon can be used to remove calcareous deposits and to soften water.

Most commercial silver cleaners also contain a sequestering agent, thiourea $(NH_2)_2CS$, together with hydrochloric acid and a detergent to cut through grease. It has been suggested that the cleaner acts by the acid first dissolving the black silver sulphide tarnish and then the silver ions forming a complex with the thiourea molecules.

D Catalysts

catalyst

Some reactions will only take place if there is a catalyst present. A **catalyst** is a substance which accelerates the rate of a reaction but is not consumed during the reaction (although it may be physically changed). Usually only small amounts of catalyst are necessary to bring about large changes in the rate of reaction.

One of the reactant molecules combines loosely with the catalyst. This combination reacts with the other reactant to form products, releasing the catalyst for further reaction.

As the reaction rate is faster in the presence of a catalyst, the activation energy of reaction must be lower. This is probably due to the entropy effect. A catalyst can provide a highly ordered environment in which the reacting molecules are brought together in specific orientations. Thus the randomness of free collisions is removed.

Because the molecules can be specifically orientated by the catalyst, a catalysed reaction may lead to products different from those due to a thermally induced reaction.

D1 Enzymes – Nature's catalysts

Organic compounds can almost invariably be destroyed in fierce enough conditions of high temperature and/or strongly oxidising atmosphere. However, you can often achieve the same ends in conditions far less hazardous to the objects being conserved by using

enzymes

enzymes. Enzymes are naturally occurring protein molecules which are found in living tissues, catalysing and directing the chemical reactions upon which life depends. Since they work within living tissue, modest temperatures and near neutral solutions obviously suffice for their action, and they are consequently useful in conservation work. Of particular importance are those enzymes concerned with the digestion of food.

Whenever an object made of organic materials decays through the actions of bacteria, fungi or moulds, it is being broken down chemically by digestive enzymes exuded by the organism. The object is being "used" as food. It is as well to remember this when enzymes are to be used as cleaning agents, as they may cause damage to the object.

Some of the particularly stubborn stains such as bloodstains, food marks or those blemishes left by mildew are also proteins and are susceptible to attack by appropriate enzymes. These are called

proteinases

proteinases, the suffix *-ase* commonly indicating an enzyme. To understand the action of enzymes and to see one example of their use, we need to turn our attention to the structural features of proteins.

amino-acids

Proteins are giant molecules made up of parts called **amino-acids**, which have the characteristic structural features shown in Figure 7.7. Just 21 of the amino-acids are the basic building blocks of all the proteins found in living matter, differing only in the complexity of their side branches. As you know, the $-NH_2$ group confers basic

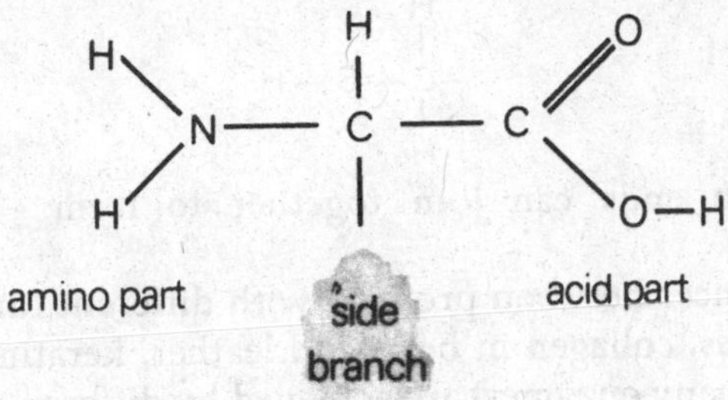

$$\mathrm{H_2N{-}CH(R){-}C({=}O){-}O{-}H}$$

amino part — side branch — acid part

Figure 7.7 *The characteristic features of amino-acids.*

properties, while $-COOH$ determines an acid nature. The $-NH_2$ and $-COOH$ groups of adjacent molecules can react:

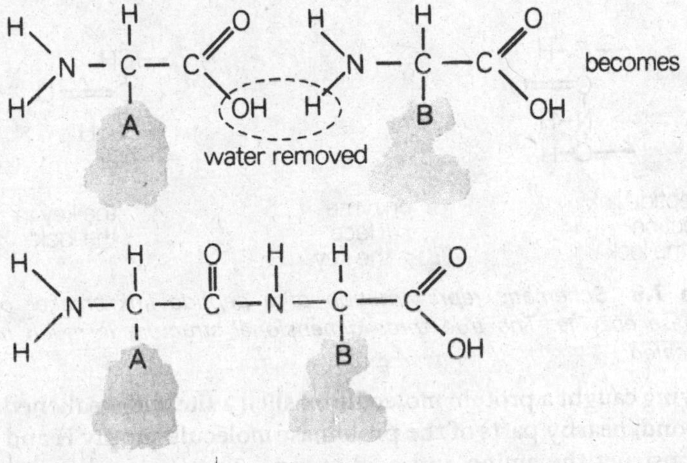

$$\mathrm{H_2N{-}CH(A){-}C({=}O){-}OH} \quad + \quad \mathrm{H{-}NH{-}CH(B){-}C({=}O){-}OH} \quad \text{becomes}$$

water removed

$$\mathrm{H_2N{-}CH(A){-}C({=}O){-}NH{-}CH(B){-}C({=}O){-}OH}$$

Figure 7.8 *Mechanism of union of amino-acids to produce proteins.*

The resulting molecule still contains both an $-NH_2$ and a $-COOH$ and so similar combinations can happen again and again. As a result, a long chain or **polymer** molecule can form, comprising the backbone:

polymer

$$-\mathrm{CH}-\mathrm{CO}-\mathrm{NH}-\mathrm{CH}-\mathrm{CO}-\mathrm{NH}-\text{etc.}$$

with various side groups on the CH carbons. The characteristic point of connection:

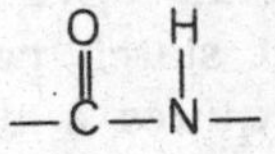

$$-\mathrm{C}(=\mathrm{O})-\mathrm{N}(\mathrm{H})-$$

between the amino acids is called a **peptide link**.

peptide link

Protein molecules, however, rarely have long thin straggling shapes and are instead three-dimensional structures of definite shape. The transformation from a long string to a more compact shape is achieved by the chain linking to itself through the side groups. There is a particular amino acid known as cysteine which serves this function. Its side branch ends with the grouping:

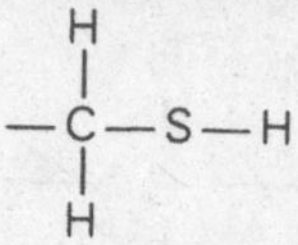

$$\begin{array}{c} H \\ | \\ -C-S-H \\ | \\ H \end{array}$$

and two such ends can join together to form $-CH_2-S-S-CH_2-$.

The differences between proteins with different functions (such as muscle fibres, collagen in bone and leather, keratin in horn and hair, and the many enzymes) are achieved by different sequences of side groups and different molecular shapes.

The enzymes in use in conservation for "digesting" stains are those which break up other proteins. These enzyme molecules have sites on their surfaces which, like a key matches a lock, match the charge distribution of the peptide link and stretch the C—N bond a little. Figure 7.9 represents this schematically:

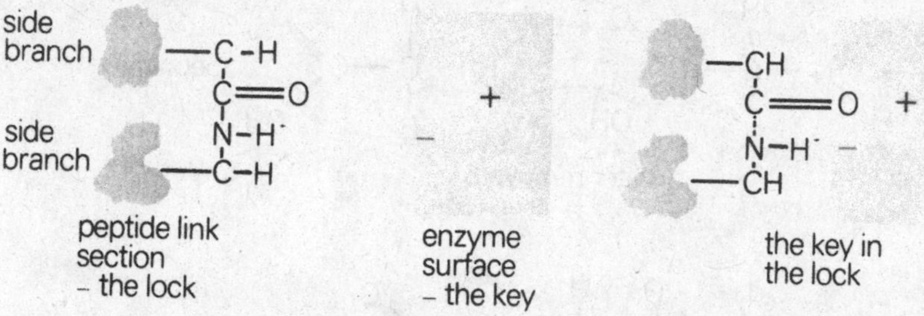

Figure 7.9 *Schematic representation of a peptide link-breaker on a proteinase enzyme. The true three-dimensional structure is much more complicated.*

Having caught a protein molecule on such a site and weakened the C–N bond, nearby parts of the proteinase molecule supply H and OH to reconstruct the amino and acid groups. The charge distribution is changed by this reaction; the broken parts of the protein molecule fall off the enzyme and the site can be used again in a similar way.

The normal requirements of animal digestion are that proteins should be broken down to their constituent amino acids which can then be dispersed throughout the body, via the bloodstream, and rebuilt into the desired proteins. Such far-reaching breakdown is not needed in order to remove most protein stains. However, even quite quick treatments may damage proteinaceous objects as well as removing stains. Residues may continue to be active. Enzymes have been successfully used to break down adhesives. Casein glues are attacked by proteinase, but starch paste, being a cellulose-like material, not a protein, requires a different enzyme, amylase. Amylase will attack paper, too, so you have to be careful where you use it. Commercial enzymes are sometimes mixtures, whose separate ingredients may digest more than you wanted.

The construction of giant molecules, polymers, from smaller fragments, which you have seen exemplified by proteins, is a subject which receives much attention in the next book of this series.

Recommended reading

Conservation science is a comparatively young discipline which has yet to develop a distinctive literature of its own. There are relatively few dedicated textbooks, and most advances in knowledge and techniques are to be found in conference preprints and proceedings and in journal articles. One consequence is that conservation students, perhaps more than others, have to ferret out the literature they require. Another is that it is not possible to present here a bibliography which precisely matches the material in this book, topic by topic.

Listed below is a selection of English language works from conservation and other disciplines which are likely to be most rewarding. They should be available in any well-equipped conservation library. The individual papers contained within them, in the journals listed and indeed in the wider international conservation literature, can be located with the help of Art and Archaeology Technical Abstracts and/or on-line via the Conservation Information Network. The latter also offers a materials database which provides technical data on conservation materials, using many of the concepts explained in this book.

International Institute for Conservation (IIC) Publications

Congress preprints:

Preprints for the IIC Rome Conference 1961 (bound volume of conference papers distributed to delegates); published as *Recent Advances in Conservation*, edited by G. Thomson, Butterworth, London, 1963.

Preprints for the IIC Delft Conference 1964 (bound volume of conference papers distributed to delegates); a fuller version in similar format appeared as *IIC 1964 Delft Conference on the Conservation of Textiles Collected Preprints*, 2nd edition, IIC, London, 1965; published as *Textile Conservation*, edited by Jentina E. Leene, Butterworth, London, 1972.

Preprints for the contributions to the London Conference on *Museum Climatology*, edited by Garry Thomson, IIC, London, 1967; revised edition May 1968.

Preprints for the contributions to the New York Conference on *Conservation of Stone and Wooden Objects*, IIC, London, 1970; second edition, edited by

Garry Thomson, published in two volumes, Volume 1 *Stone*, Volume 2 *Wooden Objects*, August 1972; subsequently reprinted as a single volume.

Conservation of Paintings and the Graphic Arts, preprints for the contributions to the Lisbon Congress 1972, IIC, London, 1972; published as *Conservation* and Restoration of Pictorial Art, edited by Norman Brommelle and Perry Smith, Butterworth, London, 1976.

Conservation in Archaeology and the Applied Arts, preprints for the contributions to the Stockholm Congress 1975, IIC, London, 1975.

Conservation of Wood in Painting and the Decorative Arts, preprints for the contributions to the Oxford Congress, edited by N. S. Brommelle, Anne Moncrieff and Perry Smith, IIC, London, 1978.

Conservation Within Historic Buildings, preprints for the contributions to the Vienna Congress, edited by N. S. Brommelle, Garry Thomson and Perry Smith, IIC, London, 1980.

Science and Technology in the Service of Conservation, preprints for the contributions to the Washington Congress, edited by N. S. Brommelle and Garry Thomson, IIC, London, 1982.

Adhesives and Consolidants, preprints for the contributions to the Paris Congress, edited by N. S. Brommelle, Elizabeth M. Pye, Perry Smith and Garry Thomson, IIC, London, 1984.

Adhésifs et Consolidants, Edition française des communications, IIC Xe Congrès International, publiée par la Section Française de l'IIC, Champs-sur-Marne, 1984.

Case Studies in the Conservation of Stone and Wall Paintings, preprints for the contributions to the Bologna Congress, edited by N. S. Brommelle and Perry Smith, IIC, London, 1986.

The Conservation of Far Eastern Art, preprints for the contributions to the Kyoto Congress, edited by John S. Mills, Perry Smith and Kazuo Yamasaki, IIC, London, 1988.

Conservation of Far Eastern Art, abstracts of the contributions to the Kyoto Congress, edited by H. Mabuchi and Perry Smith, Japanese Organizing Committee of the IIC Kyoto Congress, Tokyo, 1968.

Cleaning, Retouching and Coatings: Technology and Practice for Easel Paintings and Polychrome Sculpture, preprints for the contributions to the Brussels Congress, edited by John S. Mills and Perry Smith, IIC, London, 1990.

Cleaning, Retouching and Coatings, summaries of the posters at the Brussels Congress, IIC, London, 1990.

Conservation of the Iberian and Latin American Cultural Heritage, preprints for the contributions to the Madrid Congress, edited by H. W. M. Hodges, John S. Mills and Perry Smith, IIC, London, 1992. An abstracts booklet and a Spanish translation of the preprint volume are also planned.

Preventive Conservation: Practice Theory and Research, edited by Ray Ashok and Perry Smith, IIC, London, 1994.

ICOM Committee for Conservation

Proceedings of the following triennial meetings: (1966, 1969 and 1972 were not issued as preprints or subsequently published)

4th Triennial Meeting, Venice – 1975.
5th Triennial Meeting, Zagreb – 1978.
6th Triennial Meeting, Ottawa – 1981.
7th Triennial Meeting, Copenhagen – 1984, Diana de Froment (ed.), Paris: ICOM and the J. Paul Getty Trust.
8th Triennial Meeting, Sydney – 1987, Kirsten Grimstad (ed.), Los Angeles: ICOM CC and the Getty Conservation Institute.
9th Triennial Meeting, Dresden – 1990, J. Cliff McCawley (ed.), Los Angeles: ICOM and the Getty Conservation Institute.

United Kingdom Institute for Conservation (UKIC) Publications

Occasional Papers Series:

No. 1 *Conservation, Archaeology and Museums* (1980).
No. 2 *Microscopy in Archaeological Conservation* (1980).
No. 3 *Lead and Tin: Studies in Conservation and Technology* (1982).
No. 4 *Corrosion Inhibitors in Conservation* (1985).
No. 5 *Archaeological Bone, Antler and Ivory* (1987).
No. 6 *Restoration of Early Musical Instruments* (1987).
No. 7 *From Pinheads to Hanging Bowls: The Identification, Deterioration and Conservation of Applied Enamel and Glass Decoration on Archaeological Artifacts* (1987).
No. 8 *Evidence Preserved in Corrosion Products* (1989).
No. 9 *Conservation of Stained Glass* (1989).
No. 10 *Archaeological Textiles* (1990).

Booth, P., L. Carlyle, M. Davies, C. Leback-Sitwell, N. Kalinsky, A. Southall, V. Todd and J. Townsend, *Appearance, Opinion, Change: Evaluating the Look of Paintings*, reprint of papers given at the conference held jointly by UKIC and The Tate Gallery in January 1990, London: UKIC.

Budden, S. (ed.), *Gilding and Surface Decoration*, London: UKIC.

Entwistle, R., V. Kemp, J. Marsden and V. Todd (eds) (1992) *Life After Death: The Practical Conservation of Natural History Collections*, UKIC/Ipswich Borough Council Conference, February 1992, London: UKIC.

Fairbrass, S. and J. Hermans (eds) (1989) *Modern Art: The Restoration and Techniques of Modern Paper and Paints*, London: UKIC.

Hackney, S., J. Townsend and N. Easthaugh (eds) (1990) *Dirt and Pictures Separated*, London: UKIC.

Norman, M. and V. Todd (eds) (1991) *Storage*, preprints for the UKIC Conference, October 1991, London: UKIC.

Todd, V. (ed.) (1988) *Conservation Today*, preprints for the 30th Anniversary Conference of UKIC held in October 1988, London: UKIC.

ICCROM Publications

Masschelein-Kleiner, L. (1985) *Ancient Binding Media, Varnishes and Adhesives*, Rome: ICCROM.

Torraca, G. (1963) *Synthetic Materials used in the Conservation of Cultural Property* (4th edn 1990), Rome: ICCROM.

Torraca, G. (1975) *Solubility and Solvents for Conservation Problems* (3rd edn 1984), Rome: ICCROM.

Torraca, G. (1981) *Porous Building Materials: Materials Science for Architectural Conservation* (3rd rev. edn 1988), Rome: ICCROM.

Safety literature

Bretherick, L. (ed.) (1986) *Hazards in the Chemical Laboratory*, 4th edn, London: Royal Society of Chemistry.

Clydesdale, A. (1982) *Chemicals in Conservation: A Guide to Possible Hazards and Safe Use* (2nd edn 1987), Edinburgh: Conservation Bureau (Scottish Development Agency) and Scottish Society for Conservation and Restoration.

Howie, F. (ed.) (1987) *Safety in Museums and Galleries*, London: Butterworth with the International Journal of Museum Management.

The Health and Safety Commission (HSC) and the Health and Safety Executive publish a great deal of information which is of interest to conservators. This includes:

HSE Guidance Notes series
Health and Safety (Guidance) Series
Health and Safety (Regulations) Series.

Many are available free of charge from the HSE. Contact HSE Publications Point, St Hugh's House, Stanley Precinct, Bootle, Merseyside L20 3LZ

A full list of current HSC/E publications "Publications in Series" is published twice yearly and is available from HSE Public Enquiry Points:

Baynards House
1 Chepstow Place
Westbourne Grove
London W2 4TF

Broad Lane
Sheffield S3 7HQ

(This list applies to the UK; most other countries have their own safety organisation.)

Other publications

Allsop, Dennis and K. J. Seal (1986) *Introduction to Biodeterioration*, London: Edward Arnold.

Black, J. (ed.) (1987) *Recent Advances in the Conservation and Analysis of Artifacts*, Proceedings of the Jubilee Conservation Conference of the University of London Institute of Archaeology, London: Summer Schools Press.

Brill, T. (1980) *Light: Its Interaction with Art and Antiques*, New York: Plenum Press.

Brown, B. F., H. C. Burnett, W. T. Chase, M. Goodway, J. Kruger and M. Pourbaix (eds) (1977) *Corrosion and Metal Artifacts – a Dialogue between Conservators, Archaeologists and Corrosion Scientists*, National Bureau of Standards Special Publication 479, Washington: US Department of Commerce.

Brydson, J. A. (1989) *Plastics Materials* (5th edn), London: Butterworth.

Burns, R. M. and W. W. Bradley (1962) *Protective Coatings for Metals* (3rd edn 1967), New York: Reinhold.

Cassar, M. (1995) *Environmental Management: Guidelines for Museums and Galler* London: Museums & Galleries Commission and Routledge.

Cotterill, R. (1985) *The Cambridge Guide to the Material World*, Cambridge: Cambridge University Press.

Dana, E. S. (1991) *A Textbook of Mineralogy* (5th edn), New York: Wiley.

Eaton, L. and C. Meredith (eds) (1988) *Modern Organic Materials*, preprints for meeting held in Edinburgh, April 1988, Edinburgh: Scottish Society for Conservation and Restoration.

Feller, R. L. (1986) *Artists' Pigments: a Handbook of their History and Characteris* vol. 1, Washington: National Gallery of Art, Cambridge: Cambridge Universit Press.

Feller, R. L., N. Stolow and E. H. Jones (1985) *On Picture Varnishes and their Solvents*, Washington: National Gallery of Art.

Fleming, D., C. Paine and J. Rhodes (eds) (1992) *Social History in Museums: a Manual of Curatorship*, London: HMSO.

Franks, F. (1983) *Water* (rev. edn 1984), London: Royal Society of Chemistry.

Gettens, R. J. and G. L. Stout (1966) *Painting Materials: a Short Encyclopedia*, 2nd edn, New York: Dover Publications.

Gordon, J. E. (1968) *The New Science of Strong Materials*, Harmondsworth: Penguin.

Green, L. R. and D. Thickett (1995) 'Testing materials for use in the storage a display of antiquities. Revised methodology', *Studies in Conservation* 40.

Harley, R. D. (1980) *Artists' Pigments c. 1600–1835: a Study in Documentary Sources* (2nd edn 1982), London: Butterworth.

Hodges, H. (1964) *Artifacts – an Introduction to Early Materials and Technology* (3rd edn 1989), London: Duckworth.

Horie, C. V. (1987) *Materials for Conservation*, Sevenoaks: Butterworth.

Knell, S. (ed.) (1994) *Care of Collections*, London: Routledge.

Leigh, G. J. (1971) *Nomenclature of Inorganic Chemistry, Definitive Rules 1970* (3rd edn 1971), Oxford: Blackwell.

Long, P. and R. Levering (eds) (1979) *Paper – Art and Technology*, San Francis World Print Council.

Mayer, R. M. (1970) *The Artists' Handbook of Materials and Techniques*, ed. E. S (4th edn 1982), London: Faber & Faber.

McCrone, W. C. and J. G. Delly (1973) *The Particle Atlas*, vol. 2, Michigan: Ann Arbor Science.

McCrone, W. C., L. B. McCrone and J. G. Delly (1978) *Polarized Light Micros* Michigan: Ann Arbor Science.

Mills, J. S. and R. White (1987) *The Organic Chemistry of Museum Objects*, London: Butterworth.

Pickwood, N. (ed.) (1986–8) papers of the 10th anniversary conference, 'Ne Directions in Paper Conservation', *The Paper Conservator* vols 10–12, Oxford Worcester: Institute of Paper Conservation.

Rossotti, H. (1975) *Introducing Chemistry*, Harmondsworth: Penguin.

Sandwith, H. and S. Stainton (1993) *The National Trust Manual of Housekeeping*, Harmondsworth: Penguin.

Shields, J. (1970) *Adhesives Handbook*, London: Butterworth.

Street, A. and W. Alexander (1994) *Metals in the Service of Man* (9th edn 1989), Harmondsworth: Penguin.

Tate, J. and J. Townsend (eds) (1987) *SSCR Bulletin 9* (volume devoted to water).

Tate, J. O., N. H. Tennent and J. H. Notman (eds) (1983) *Resins in Conservation*, Proceedings of conference held in May 1982, Edinburgh: Scottish Society for Conservation and Restoration.

Thomson, G. (1978) *The Museum Environment* (2nd edn 1987), London: Butterworth in association with International Institute for Conservation.

See also various publications of:

The Open University, Science Foundation Course, Unit S102
Royal Institute of Chemistry.

Journals, newsletters and conference proceedings of the following organisations:

American Institute for Conservation (AIC)
Australian Institute for the Conservation of Cultural Material (AICCM)
Canadian Conservation Institute (CCI)
Institute of Paper Conservation (IPC)
International Institute for Conservation (IIC)
IIC–Canadian Group
Museums Association (MA)
Scottish Society for Conservation and Restoration (SSCR)
Society for the Preservation of Natural History Collections (SPNHC)
United Kingdom Institute for Conservation (UKIC)

Journals

AICCM Bulletin
Collection Forum (SPNHC)
The Conservator (UKIC)
Journal of the American Institute for Conservation
Journal of the IIC–Canadian Group
Museum Practice (MA)
The Paper Conservator (IPC)
Restaurator
SSCR Journal
Studies in Conservation (IIC)

Art and Archaeology Technical Abstracts (formerly *IIC Abstracts*), published semi-annually by the Getty Conservation Institute in association with the International Institute for Conservation of Historic and Artistic Works. AATA is an international abstracting journal for conservation.

The *List of Acquisitions* and *Subject Index* are published every two years by ICCROM commencing 1977.

The Conservation Information Network

The Conservation Information Network is an international co-operative projec initiated by the Documentation Program of the Getty Conservation Institute (based in California). The Network consists of a series of databases: bibliographic, conservation materials, conservation supplies and equipment databas These are held on a mainframe computer at the offices of the Canadian Herita Information Network in Ottawa, Canada. The databases can be accessed via th Canadian Heritage Information Network's World Wide Web Site at http://www.chin.gc.ca. The Network was launched in 1987 and there are ove 500 individuals and institutions from around the world subscribing to it. The operating partners are: AATA, Conservation Analytical Laboratory of the Smithsonian Institution, ICCROM, ICOMOS, ICOM and CCI. Further information about the Network is available from:

Conservation Information Network
Communications Canada
365 Laurier Avenue West
Journal Tower South, 12th Floor
Ottawa, Ontario
Canada K1A 0C8

Canadian Heritage Information Network
Department of Canadian Heritage
Les Terrasses de la Chaudière
15 Eddy Street, 4th Floor
Hull, Quebec
Canada K1A OM5

Within the United Kingdom, information is also available from:

The Conservation Unit
Museums & Galleries Commission
16 Queen Anne's Gate
London SW1H 9AA

Index

Photographic credit

6.11 Pye Unicam Ltd

in the United States
& Taylor Publisher Services